高等学校 电气工程及其自动化专业 应用型本科

U0676952

Introduction to Electrical Engineering

电气工程导论

主　编　魏　钢　曹政钦　王　佳

副主编　聂　玲　杨　滔　李国勇

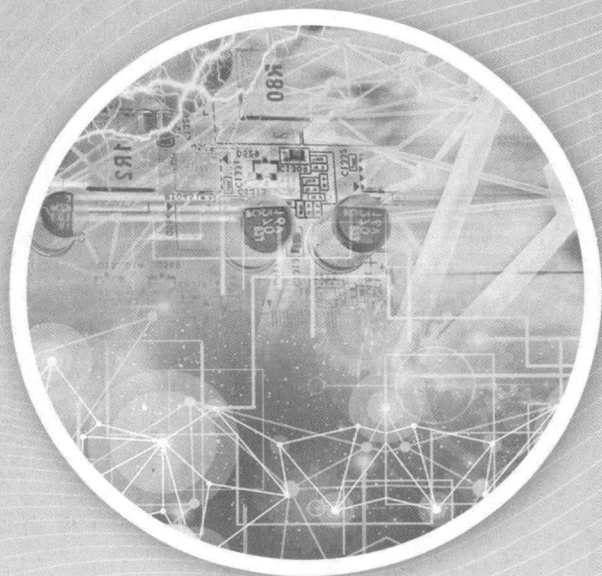

重庆大学出版社

内容提要

本书是一本电气工程相关专业学生的入门教材,共有 7 个章节,主要内容包括总论、电路原理基础、电机与电气、电力电子、电力系统及自动化、高电压与绝缘技术和智能电网,列举了电气工程技术在一些重要行业的应用案例。本书通过对电气工程学科的起源、主要研究内容、现状和未来发展趋势的介绍,让学生对电气学科有基本的认识,激发学生的学习兴趣和积极性,为后续课程学习奠定良好的基础。

本书深入浅出、内容翔实,体现了电气工程的基础性、系统性、前沿性和理论性,反映了本领域的新动态、新观点,适合本科类院校电气类专业高校本科生,尤其是来自英语国家的国际学生,也可作为英语国家非电气行业初学者、部分电气从业人员的培训教材和学习参考书。

图书在版编目(CIP)数据

电气工程导论 = Introduction to Electrical
Engineering:英文/魏钢,曹政钦,王佳主编. -- 重
庆:重庆大学出版社,2022.3
高等学校电气工程及其自动化专业应用型本科系列教材
ISBN 978-7-5689-2435-1

Ⅰ.①电… Ⅱ.①魏… ②曹… ③王… Ⅲ.①电气工
程—高等学校—教材—英文 Ⅳ.①TM

中国版本图书馆 CIP 数据核字(2021)第 130722 号

Introduction to Electrical Engineering
电气工程导论
DIANQI GONGCHENG DAOLUN

主 编 魏 钢 曹政钦 王 佳
副主编 聂 玲 杨 滔 李国勇
策划编辑:杨粮菊
责任编辑:范 琪 荀荟羽   版式设计:杨粮菊
责任校对:关德强   责任印制:张 策

*

重庆大学出版社出版发行
出版人:饶帮华
社址:重庆市沙坪坝区大学城西路 21 号
邮编:401331
电话:(023)88617190 88617185(中小学)
传真:(023)88617186 88617166
网址:http://www.cqup.com.cn
邮箱:fxk@ cqup.com.cn(营销中心)
全国新华书店经销
重庆天旭印务有限责任公司印刷

*

开本:787mm×1092mm 1/16 印张:8.5 字数:280 千
2022 年 3 月第 1 版   2022 年 3 月第 1 次印刷
ISBN 978-7-5689-2435-1   定价:45.00 元

Preface

The electrical engineering covers a wide range of fields, such as electricity, electronics and electromagnetism. Since the commercialization of the telegraph and power supply in the late 19th century, the electrical engineering has been identified as a discipline for the first time. Now it has developed into a discipline covering electricity, electronics, control system, signal processing and telecommunication.

As an introductory course for freshmen, this book discusses the latest progress of electrical engineering, involving computer, power electronics, artificial intelligence, etc., through the overall study and introduction to related courses of electrical engineering on the basis of the basic principles and theories, so as to stimulate students' interest and enthusiasm in learning and lay a good foundation for the follow-up courses.

This book is mainly compiled by Professor Wei Gang, lecturer Nie Ling, PH. D Cao Zhengqin from Chongqing University of Science and Technology, lecturer Wang Jia from Chongqing Medical University, senior engineer Yang Tao from State Grid Chongqing Electric Power Company, and senior engineer Li Guoyong from Hunnan Electric Power Company Limited. In addition, this book is divided into seven chapters, namely general introduction, circuit principle basis, motor and electrical appliances, power electronics, power system and its automation, high voltage and insulation technology, and smart grid. The first chapter (compiled by Wang Jia and Wei Gang) defines the concept of electrical engineering, introduces the brief history of electrical engineering, displays the system of units and the commonly used computer software in this field, and shows the special features of this course. The second chapter (compiled by Cao Zhengqin) defines the fundamentals of electric circuits, introduces the circuit principle, including the Ohm's law and Kirchhoff's law, and describes the electric power and sign convention, basic analytical methods and Sinusoidal AC analysisand three-phase circuit. The third chapter (compiled by Wei Gang) introduces the classification of electric machine, the alternator, generator, transformer, motor, special motor, and high/low-voltage apparatus. The fourth chapter (compiled by Nie Ling and Wang Jia) introduces the generation and development of power electronic technology, power electronic device and power electronic converter technology. The fifth chapter (compiled by Wei Gang) introduces the electric power industry, composition of power system, power plants, transmission and distribution system, substation, power supply and utility technology, and protection and remote control system. The sixth chapter (compiled by Cao Zhengqin and Yang Tao) defines the main contents of the course, and introduces high voltage insulation technology, including the source of high-voltage disciplines, dielectric insulation, measurement of high voltages, and overvoltage protection and insulation coordination. The seventh chapter (compiled by Wang Jia and Li Guoyong) introduces the background and features of the smart grid, and displays the technology and modelling of smart grid, and presents the deployments and potential deployment. This book is reviewed by Wei Gang and Wang Jia.

With plain language of easy comprehension and detailed content, this book represents the fundamental, systematic, pioneering and theoretical nature of electrical engineering, and it introduces the new trends and views in this field.

This book is suitable for undergraduates majoring in electrical engineering, especially for overseas students from English-speaking countries, and for the readers who wish to get the initial training in non-electrical industry and electrical practitioners in English speaking countries as well.

Due to the inadequate experience of the authors, the structure and content of this book may be not perfectly reasonable, and there may be some inevitable errors and omissions existing in this book. Any suggestion or criticism will be heartily appreciated.

Editor
Summer 2020

前　言

电气工程领域广泛,涉及电力、电子和电磁学的研究与应用。从 19 世纪末电报和电力供应商业化后,电气工程首次被确定为一门学科,到现在它发展成为一门涵盖电力、电子、控制系统、信号处理和电信的学科。

作为本科新生的入门课程教材,本书通过对电气工程学科的起源、主要研究内容、现状和未来发展趋势的介绍,让学生对电气学科有基本的认识,激发学生的学习兴趣和积极性,为后续课程学习奠定良好的基础。

本书主要由重庆科技学院魏钢教授、曹政钦博士、聂玲讲师、重庆医科大学王佳讲师、国网重庆市电力公司杨滔高级工程师、国网湖南省电力有限公司李国勇高级工程师编写。本书分为 7 个章节,分别是总论、电路原理基础、电机与电器设备、电力电子、电力系统及其自动化、高电压与绝缘技术和智能电网。

第一章总论为本书的开篇章节,界定了电气工程这一概念,介绍了电气工程发展历程和电气工程中常用的计算机软件。该章由王佳、魏钢编写。第二章电路原理基础,介绍了欧姆定律、基尔霍夫定律以及电力与标志公约,阐述了电路元件、基本分析方法以及正弦交流电路分析和三相电路的相关知识。该章由曹政钦编写。第三章电机与电器设备,介绍了电动机的分类,并围绕交流发电机、变压器、电动机、专用电动机和高/低压电器进行了逐一讲解。该章由魏钢编写。第四章电力电子,介绍了电力电子技术的产生和发展历史,并分析了电力电子装置、电力电子变换器技术以及电气传动技术。该章由聂玲、王佳编写。第五章电力系统及其自动化,介绍了电力系统构成、发电厂、电力传输和分配系统、变电站、供电和公用技术、保护和远程控制系统等内容。该章由魏钢编写。第六章高电压与绝缘技术,介绍了电介质绝缘、高电压试验技术、过电压与绝缘配合等内容。该章由曹政钦、杨滔编写。第七章智能电网,介绍了智能电网的特点,并就智能电网研究、部署进行了探讨。该章由王佳、李国勇编写。全书由魏钢和王佳审核。

本书深入浅出、内容翔实,体现了电气工程的基础性、系统性、前沿性和理论性,反映了本领域的新动态、新观点,适合本科类院校电气类专业本科生,尤其是来自英语国家的国际学生学习,也可作为英语国家非电气行业初学者、部分电气从业人员的培训教材和学习参考书。

由于作者能力有限,本书的结构体系和内容取舍不一定完全合理,书中难免存在疏漏之处,恳请读者批评指正。

编　者
2021 年夏

Contents

Chapter *1*
Introduction

This chapter is to provide the newcomer with a general view of the different specialties in electrical engineering and a better understanding of the aim and organization of the book.

This chapter begins by defining electrical engineering and its various branches, presenting some interactions among them, and illustrating how electrical engineering is closely connected to many other engineering disciplines by means of a practical example. A brief historical perspective is also provided to outline the growth and development of this relatively young engineering specialty. Next, the fundamental physical quantities and the system of units are explained to set the stage for the chapters that follow. Finally, the organization of this book is elaborated to give the students, as well as the teachers, the continuity of the target development throughout the chapters 1 and 2.

1.1   Electrical engineering

Figure 1.1   William Gilbert and his book

William Gilbert, a British scientist, published a book *On the Magnet* in 1600, as shown in Figure 1.2, which systematically discussed the magnetism of the earth and considered the earth as a large magnet. The magnetic dip angle can be used to determine the latitude of the earth. On the basis of the Greek ($\eta \lambda \varepsilon \kappa \tau \rho o \nu$) and Latin (electrum), the word "Electricity" was created.

Figure 1.2　Discharge phenomenon

Electricity is a natural phenomenon caused by the movement of electric charge. Electricity is an attribute of repulsion and attraction between subatomic particles, such as electrons and protons. As one of the four basic interactions in nature, electricity is a general term. It is a physical phenomenon caused by a static or moving charge. In nature, the mechanism of electricity brings many common effects, such as lightning, friction electrification, electrostatic induction, electromagnetic induction and so on.

Electrical engineering (abbreviated EE), sometimes referred to as electrical and electronic engineering, is a field of engineering involving the research and application of electricity, electronics and electromagnetism. Electrical engineering was first identified as a discipline in the late 19th century, after the commercialization of telegraphy and power supply. It has now evolved into a range of disciplines covering electricity, electronics, control systems, signal processing and telecommunications.

As one of the core subjects in the field of modern science and technology, electrical engineering is also an indispensable discipline in the field of high-tech electrical engineering.

For example, it is the great progress of electronic technology that has promoted the arrival of the information age based on computer networks to change the mode of human life and work. The bright future of electrical engineering has leaded to a high employment rate of students recently.

Electrical engineering in the United States also includes electronic engineering. In other countries, electrical engineering is considered as large-scale power systems, such as power transmission and motor control, while electronic engineering is considered as small-scale electronic systems, including computers and integrated circuits.

The traditional electrical engineering is defined as "the disciplines involved in creating electrical and electronic systems". That is a very broad definition, but with the rapid development of science and technology, the concept of electrical engineering in the 21st century has gone far beyond the definition defined above.

With the progressive development of basic theory research and science and technology, the proposal of new design philosophy and design method will have a great influence on the development trend of the electric engineering discipline in the next years. ① The advancement of information technology will have a decisive influence on the development of electric engineering discipline. ② The widening of the intersections between electrical engineering disciplines and physical science will bring new opportunities for the development of electrical engineering disciplines. ③ The fast developing new technology and method will provide a more scientific and technical solution for the electric engineering disciplines.

In addition, electrical engineers are usually connected with the use of electricity to transmit energy, while electronic engineers are more interested in information delivery.

(a) IEEE

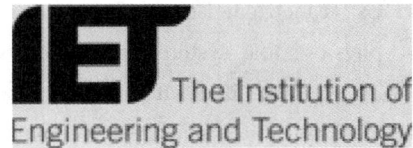
(b) IET

(c) CSEE

Figure 1.3　Logos of various electrical society

Here are some famous academic groups in the electrical field, as shown in Figure 1.3.

IEEE (Institute of Electrical and Electronics Engineers) is the world's largest technical professional organization dedicated to advancing technology for the benefit of humanity.

Website: www. IEEE. org

IEEE and its members inspire a global community to innovate for a better future through its over 423,000 members in over 160 countries, and its highly cited publications, conferences, technology standards, and professional and educational activities. IEEE is the trusted "voice" for engineering, computing, and technology information around the globe.

Its headquarter is in New York, the United States.

IET (Institution of Engineering and Technology) is one of the world's largest engineering institutions with over 168,000 members in 150 countries. It is also the most multi-disciplinary to reflect the increasingly diverse nature of engineering in the 21st century.

Website: www. IET. org

The IET is working for a better world by inspiring, informing and influencing members, engineers and technicians, and all those who are touched by, or touching, the work of engineers.

Its headquarters are in Stevenage near London, England.

CSEE (Chinese Society for Electrical Engineering) is a non-profit academic social body formed voluntarily by Chinese electrical engineering scientists and technicians registered in accordance with the law. It is an important social force in the cause of China's electrical engineering science and technology.

Website:www. CSEE. org. cn

CES (China Electrotechnical Society)

Website:www. CES. org. cn

It was founded in 1934 with its working headquarter in Beijing.

Now let's take a look at the concept of engineer.

What is engineer? Engineers can do anything.

Invent—Develop a new product, system, or process that has never existed before.

Innovate—Improve an existing technological product, system, or method.

What is **EEs**? Electrical Engineers are inventors and innovators who apply knowledge of signals, circuits, physics, and systems to develop technologies to improve people's lives.

The following example illustrates the pervasive presence of electrical, electronic, and electromechanical devices and systems in a very common application: the automobile.

EXAMPLE:Electrical Systems in a Passenger Automobile.

A familiar example illustrates how the seemingly disparate specialties of electrical engineering actually interact to permit the operation of a very familiar engineering system: the automobile. Figure 1.4 presents a view of electrical engineering systems in a modern automobile. Even in older vehicles, an electric circuit plays a very important role in the overall operation. An inductor coil generates a sufficiently high voltage to allow a spark to form across the spark plug gap, and to ignite the air and fuel mixture; the coil is supplied by a DC voltage provided by a lead-acid battery. In addition to providing the energy for the ignition circuits, the battery also supplies power to many other electrical components, such as lights, the windshield wipers, and the radio. Electric power is carried from the battery to all of these components by means of a wire harness, which constitutes a rather elaborate electrical circuit. In recent years, the conventional electrical ignition system has been supplanted by electronic ignition; that is, solid-state electronic devices called transistors have replaced the traditional breaker points. The advantage of transistorized ignition systems over the conventional mechanical ones is their greater reliability, easy control, and life span (mechanical breaker points are subject to wear). The on-board radio receives electromagnetic waves by means of the antenna, and decodes the communication signals to reproduce sounds and speech of remote origin; other common communication systems that exploit electromagnetics are CB radios and cellular phones. But this is not all! The battery is, in effect, a self-contained 12-VDC electric power system, providing energy for all of the aforementioned functions. In order to have a longer service life of the battery, a charging system, composed of an alternator and of power electronic devices, is placed in every automobile. The alternator is an electric machine, as the motors that drive the power mirrors, power windows, power seats, and other convenience features found in luxury cars. Incidentally, the loudspeakers are also electric machines!

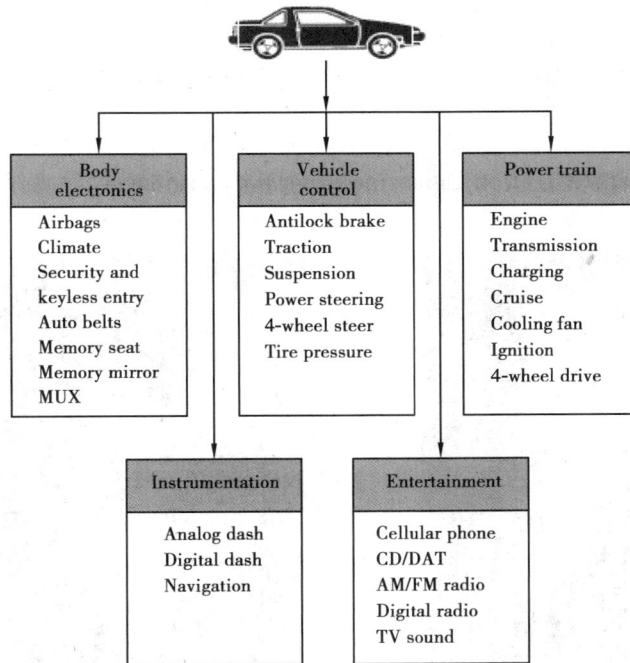

Body electronics	Vehicle control	Power train
Airbags	Antilock brake	Engine
Climate	Traction	Transmission
Security and	Suspension	Charging
keyless entry	Power steering	Cruise
Auto belts	4-wheel steer	Cooling fan
Memory seat	Tire pressure	Ignition
Memory mirror		4-wheel drive
MUX		

Instrumentation	Entertainment
Analog dash	Cellular phone
Digital dash	CD/DAT
Navigation	AM/FM radio
	Digital radio
	TV sound

Figure 1.4   Electrical engineering systems in the automobile

The list does not end here, though. In fact, some of the more interesting applications of electrical engineering to the automobile have not been discussed yet. In terms of computer systems, you are certainly aware that in the last two decades, environmental concerns related to exhaust emissions from automobiles have led to the introduction of sophisticated engine emission control systems. The heart of such control systems is a type of computer called a microprocessor. The microprocessor receives signals from devices (called sensors) that measure relevant variables such as the engine speed, the concentration of oxygen in the exhaust gases, the position of the throttle valve (i. e., the driver's demand for engine power) and the amount of air aspirated by the engine, and subsequently computes the optimal amount of fuel and the correct timing of the spark to result in the cleanest combustion possible under the circumstances. The measurement of the aforementioned variables falls under the heading of instrumentation, and the interconnection between the sensors and the microprocessor is usually made up of digital circuits. Finally, as the presence of on-board computers becomes more pervasive in areas such as antilock braking, electronically controlled suspensions, four-wheel steering systems, and electronic cruise control communications among the various on-board computers will have to occur at faster rates. Some day in the not-so-distant future, these communications may occur in a fiber optic network, and electro-optics will replace the conventional wire harness. It should be noted that electro-optics is already present in some of the more advanced displays that are part of an automotive instrumentation system.

1.2　Brief history of electrical engineering

The historical evolution of electrical engineering can be attributed, in part, to the work and discovery of the scientists as follows.

Figure 1.5　Charles A. Coulomb

Figure 1.6　Andre Marie Amper

William Gilbert (1540—1603), English physician, founder of magneticscience, published De Magnete, a treatise on magnetism, in 1600.

Charles A. Coulomb (1736—1806), French engineer and physicist, published the Laws of Electrostatics in Seven Memoirs to the French Academy of Science between 1785 and 1791. His name is associated with the unit of charge.

James Watt (1736—1819), English inventor, developed the steam engine. His name is used to represent the unit of power.

Alessandro Volta (1745—1827), Italian physicist, discovered the electric pile. The unit of electric potential and the alternate name of this quantity (voltage) are named after him.

Hans Christian Oersted (1777—1851), Danish physicist, discovered the connection between electricity and magnetism in 1820. The unit of magnetic field strength is named after him.

Andre Marie Ampere (1775—1836), French mathematician, chemist, and physicist, experimentally quantified the relationship between electric current and the magnetic field. His works were summarized in a treatise published in 1827. The unit of electric current is named after him.

Georg Simon Ohm (1789—1854), German mathematician, investigated the relationship between voltage and current and quantified the phenomenon of resistance. His first research were published in 1827. His name is used to represent the unit of resistance.

Michael Faraday (1791—1867), English experimenter, demonstrated electromagnetic induction in 1831. His invention such as electrical transformer and electromagnetic generator marked the beginning of the age of electric power. His name is associated with the unit of capacitance.

Figure 1.7   Georg Simon Ohm

Figure 1.8   Michael Faraday

Joseph Henry (1797—1878), American physicist, discovered self-induction around 1831, and his name has been designated to represent the unit of inductance. He had also developed the essential structure of the telegraph, which was later perfected by Samuel F. B. Morse.

Carl Friedrich Gauss (1777—1855), German mathematician, and **Wilhelm Eduard Weber** (1804—1891), German physicist, published a treatise in 1833 describing the measurement of the earth's magnetic field. The gauss is a unit of magnetic field strength, while the weber is a unit of magnetic flux.

Figure 1.9   James Clerk Maxwell

Figure 1.10   Nikola Tesla

James Clerk Maxwell (1831—1879), Scottish physicist, discovered the electromagnetic theory of light and the laws of electrodynamics. The modern theory of electromagnetics is entirely founded upon Maxwell's equations.

Ernst Werner Siemens (1816—1892) and Wilhelm Siemens (1823—1883), German inventors and engineers, contributed to the invention and development of electric machines, as well as to perfecting electrical science. The modern unit of conductance is named after them.

Heinrich Rudolph Hertz (1857—1894), German scientist and experimenter, discovered the nature of electromagnetic waves and published his findings in 1888. His name is associated with the unit of frequency.

Nikola Tesla (1856—1943), Croatian inventor, emigrated to the United States in 1884. He invented polyphase electric power systems and the induction motor and pioneered modern AC electric power systems. His name is used to represent the unit of magnetic flux density.

1.3　System of units

This book employs the International System of Units (also called SI) which are commonly adhered to by virtually all engineering professional societies. This section summarizes SI units serve as a useful reference in reading the book. SI units have six fundamental quantities, listed in Table 1.1. All other units may be derived in terms of the fundamental units of Table 1.2. Since, in practice, one often needs to describe quantities that occur in large multiples or small fractions of a unit, standard prefixes are used to denote powers of 10 of SI (and derived) units. These prefixes are listed in Table 1.3. It should be noted that, in general, engineering units are expressed in powers of 10 that are multiples of 3.

Table 1.1　SI units

Quantity	Unit	Symbol
Length	Meter	m
Mass	Kilogram	kg
Time	Second	s
Electric current	Ampere	A
Temperature	Kelvin	K
Luminous intensity	Candela	cd

Table 1.2　Standard prefixes

Prefix	Symbol	Power
atto	a	10^{-18}
femto	f	10^{-15}
pico	p	10^{-12}
nano	n	10^{-9}
micro	μ	10^{-6}
milli	m	10^{-3}

Table 1.2(continued)

Prefix	Symbol	Power
centi	c	10^{-2}
deci	d	10^{-1}
deka	da	10
kilo	k	10^{3}
mega	M	10^{6}
giga	G	10^{9}
tera	T	10^{12}

1.4　Special features of this course

At present, in the electrical engineering and automation specialty of most universities, there are 5 second-class disciplines at the first-level sub-discipline of electric engineering, namely, electrical machines and electric apparatus, power system and automation, high-voltage and insulation technology, power electronics and electric power drives, theory and new technology of EE.

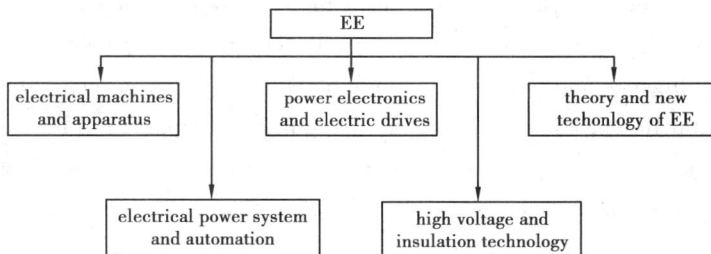

Figure 1.11　Constitution of EE discipline in China

The typical curriculum of an undergraduate electrical engineering student includes the subjects listed in Table 1.3.

Table 1.3　The typical curriculum of electrical engineering disciplines

Circuit principle
Electromagnetism
Electric machines
Computer systems
Electronic technology
Electric power systems analysis
Power supply and distribution system

Table 1.3(continued)

High-voltage technology
Power system relay protection
Control systems
Automatic device principle

The electrical engineering discipline focuses on theory research and engineering practice to strengthen theoretical basis and broaden professional knowledge. With the technological progress of electrical engineering and the improvement of the automation level, the technical talents of electrical engineering discipline are supposed to master information technology, automation technology and computer technology.

Therefore, the education of the discipline is characterized by the combination of strong and weak electricity, power technology and electronic technology, components and systems, and computer software and hardware.

Juniors are cultivated on the general education and fundamental courses, while seniors are supposed to choose specialty or professional orientation according to their interests mainly on the fundamental course and professional disciplines, so as to facilitate interdisplinary study each other, enlarge students' learning autonomy, and arouse students' enthusiasm, initiative and creativity.

To cultivate the senior professional talents engaging in the development, manufacture, operation and management related with electrical engineering with the qualities as follows: all-round development of morality, intelligence and physique, solid foundation, wide knowledge, innovative spirit and practical ability, unification of knowledge, ability and quality, adaptation of the need for technology and production development of enterprises, including electrical equipment plant, power plants, electrical power company, institute and design institute.

1.5　The common computer softwares in electrical engineering

With the development of information technology, computers have been widely used in electrical engineering. The following is a brief introduction to some digital computer simulation software widely used in electrical engineering.

1.5.1　CAD

CAD (Computer Aided Design), a numerically controlled program that allowed designers to draw simple lines with a computer, was invented in 1950. At that time, computers were the size of a room and very expensive, so this type of program was not widely available.

In the 1970s, as the cost of minicomputer fallen, interactive mapping systems became widely used in American industry. In the 1980s, with the rapid development of PC, companies specializing in CAD system emerged.

Powerful, convenient to use and reasonable in price, CAD drawing software has been the most popular CAD software package in the field of computer-aided design. It has been widely used in many industries such as machinery, architecture, home furnishing, textile, etc. at home and abroad.

1.5.2 **MATLAB**

MATLAB is a commercial mathematics software produced by MathWorks. It is an advanced technical computing language and interactive environment for algorithm development, data visualization, data analysis and numerical calculation, mainly including MATLAB and Simulink.

MATLAB is a combination of the two words matrix&laboratory, which means matrix factory (matrix lab). It is a high-tech computing environment for scientific computing, visualization and interactive programming released by MathWorks. It integrates numerical analysis, matrix calculation, scientific data visualization and nonlinear dynamic system of modeling and simulation, and many other strong function integration in the Windows environment, for scientific research, engineering design and effective numerical calculation, etc. It represents the current international scientific computing software advanced level by providing a comprehensive solution and getting rid of the traditional non-interactive programming language (such as C, Fortran) edit mode to a great extent. MATLAB can carry out matrix calculation, draw functions and data, implement algorithms, create user interfaces and connect programs of other programming languages, and it has been mainly applied in the fields of engineering calculation, control design, signal processing and communication, image processing, signal detection, financial modeling design and analysis.

1.5.3 **EMTP**

EMTP (Electro-Magnetic Transient Program) is a simulation software for electromagnetic Transient analysis of power systems. In order to simulate the HVDC system, the program has increased the capability of simulating diode, thyristor and other switching devices. Like the SPICE program, there are several EMTP versions for personal computers, such as Micro Tran and ATP. All versions of the program have most of the functions of the original EMTP version of BPA (Bonneville Power Administration).

EMTP, based on the trapezoidal integral rule, used the adjoin model as the dynamic element, established the equation with the node method, and solved the algebraic equation with the sparse matrix and LU factor decomposition method. The integral step is longer than what the user specified, and it remains unchanged throughout the simulation. Different from the switch in SPICE which is represented by non-linear resistance, the switch in EMTP is expressed as: open when disconnected; short circuit and two associated nodes being combined into one when connected. EMTP has a module called TACS and a module called MODEL to simulate the controller.

1.5.4 **PSPICE**

PSPICE is a general circuit analysis program developed from SPICE (Simulation Program with Integrated Circuit Emphasis) for microcomputer series. Developed in 1972 by a computer-aided design team at the university of California, Berkeley, using FORTRAN language, it has been mainly used in computer-aided design of large-scale integrated circuits.

Launched in 1975, SPICE's official SPICE 2G ran on at least minicomputers. In 1985, the University of California, Berkeley, adapted SPICE software using C and launched by MICROSIM. In 1988, SPICE was established as America's national industrial standard. At the same time, various commercial analog circuit simulation software have done a lot of practical work on the basis of SPICE, which helps SPICE become the most popular electronic circuit simulation software.

Chapter **2**

Fundamentals of Electric Circuits

2.1 Definitions

In this section, some variables and concepts that appear in the follouing chapters are defined. First, voltage and current sources and various load are defined; next, the concepts of branch, node, loop and mesh which form the basis of circuit analysis are explained.

2.1.1 Ideal voltage sources

An ideal voltage source is an electric device that generates a prescribed voltage at its terminals. The ability of an ideal voltage source to generate its output voltage is not affected by the current which flows out to the load or other circuit elements. Figure 2.1 depicts the symbol used to represent the ideal voltage sources.

2.1.2 Ideal current sources

An ideal current source is a device that generates a prescribed current independent of the connected circuit. Therefore, it must be able to generate an arbitrary voltage across its terminals. Figure 2.2 depicts the symbol used to represent ideal current sources.

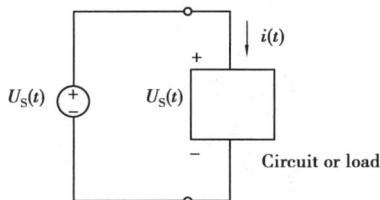

Figure 2.1   Ideal voltage sources

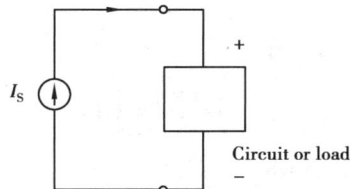

Figure 2.2   Ideal current sources

2.1.3 Branch

A branch is any portion of a circuit with two terminals connected to it. A branch may consist of one or more circuit elements (Figure 2.3). In practice, any circuit element with two terminals connected to it is a branch.

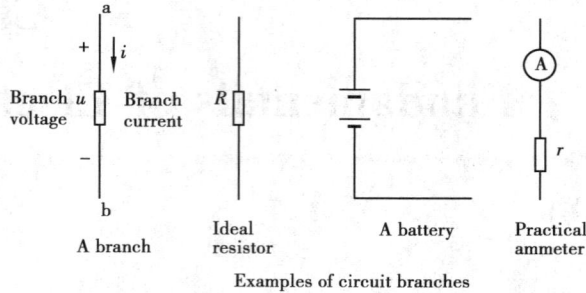

Examples of circuit branches

Figure 2.3   Examples of circuit branches

2.1.4 Node

A node is the junction of two or more branches (one often refers to the junction of only two branches as a trivial node), as shown in Figure 2.4. Any connection that can be accomplished by soldering various terminals together is a node. It is very important to identify nodes properly in the analysis of electrical networks.

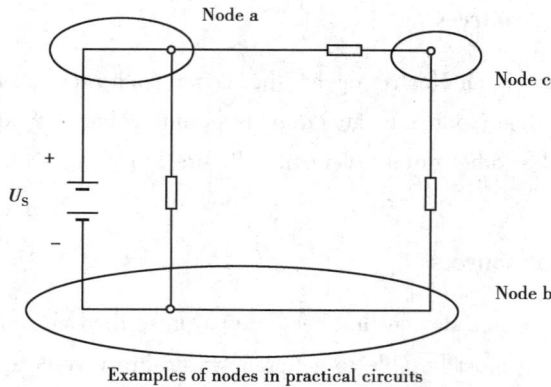

Examples of nodes in practical circuits

Figure 2.4   Examples of nodes in practical circuits

2.1.5 Loop

A loop is any closed connection of branches.

2.1.6 Mesh

A mesh is a loop that does not contain other loops. Meshes are important aid to certain analysis methods. In Figure 2.5, loop 1 and loop 2 are meshes, but loop 3 is not a mesh, because it encircles both loop 1 and loop 2. Figure 2.6 illustrates how meshes are simpler to visualize in complex networks than loops are.

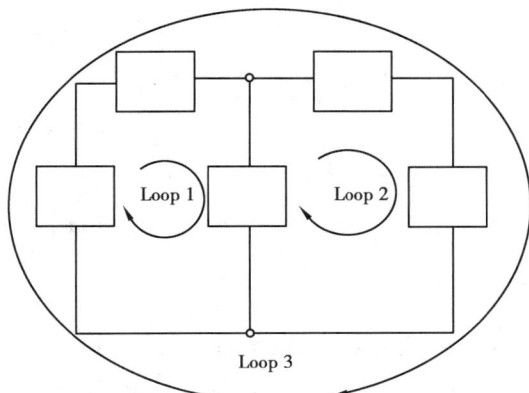

Figure 2.5   Examples of loops

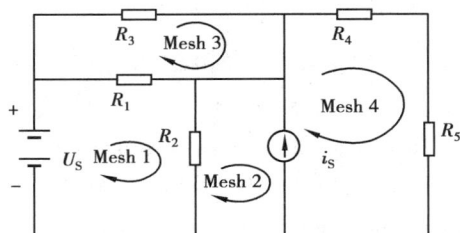

Figure 2.6   Examples of meshes

2.2   Ohm's Law

The circuit element used to model the current-resisting behavior of a material is the resistor which is the simplest passive element.

Georg Simon Ohm (1787—1854), a German physicist, is credited with formulating the current-voltage relationship for a resistor based on experiments performed in 1826. This current-voltage relationship is known as Ohm's Law.

Ohm's Law states that the voltage across a resistor is directly proportional to the current flowing through the resistor. The constant of proportionality is the resistance value of the resistor in ohms. The circuit symbol for the resistor is shown in Figure 2.7.

For the current and voltage shown, Ohm's law is

$$u(t) = Ri(t) \qquad (2.1)$$

where $R \geqslant 0$ is the resistance in ohms.

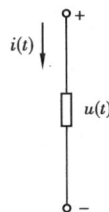

Figure 2.7   Symbol for the resistor

Rearranging Eq. (2.1) into the form $R = u(t)/i(t)$, 1 ohm = 1 V/A. The symbol used to represent the ohm is the capital Greek omega (Ω) .

Since R is a constant, Eq. (2.1) is the equation of a straight line. For this reason, the resistor is called a linear resistor. A graph of $u(t)$ versus $i(t)$ is shown in Figure 2.8, which is a line passing through the origin with a slope of R. Obviously, a straight line is the only graph possible for which the ratio of $u(t)$ to $i(t)$ is a constant for all $i(t)$.

Resistors whose resistances do not remain constant for different terminal currents are known as nonlinear resistors. For such a resistor, the resistance is a function of the current flowing in the device. A simple example of a nonlinear resistor is an incandescent lamp. A typical voltage-current characteristic for this device is shown in Figure 2.9, where the graph is no longer a straight line. Since R is not a constant, the analysis of a circuit containing nonlinear resistor is more difficult.

15

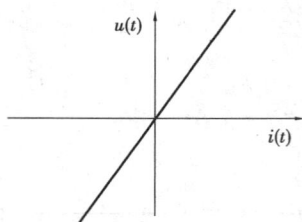

Figure 2.8　Typical voltage-current characteristics for a linear resistor

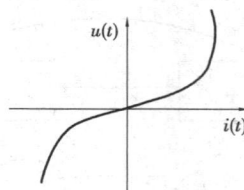

Figure 2.9　Typical voltage-current characteristics for a nonlinear resistor

In reality, all practical resistors are nonlinear because the electrical characteristics of all conductors are affected by environmental factors such as temperature. Many materials; however, are closely approximate an ideal linear resistor over a desired operating region. These types of elements are simply referred to as resistors.

Since the value of R can range from zero to infinity, it is important to consider the two extreme possible values of R. $R = 0$ is called a short circuit, as shown in Figure 2.10(a). For a short circuit,

$$u = iR = 0 \tag{2.2}$$

the voltage is zero, but the current could be any value. In practice, a short circuit is usually a connecting wire assumed a perfect conductor. Thus, a short circuit is a circuit element with resistance approaching zero.

Similarly, $R = \infty$ is known as an open circuit, as shown in Figure 2.10 (b). For an open circuit, the current is zero, but the voltage could be any value.

$$i = \lim_{R \to \infty} \frac{u}{R} = 0 \tag{2.3}$$

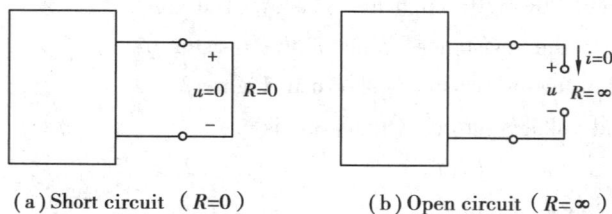

(a)Short circuit (R=0)　　(b)Open circuit (R=∞)

Figure 2.10　Short circuit and open circuit

Thus, an open circuit is a circuit element with resistance approaching infinity.

Another important quantity which is useful in circuit analysis is known as conductance, defined by

$$G = \frac{1}{R} = \frac{i}{u} \tag{2.4}$$

The conductance is a measurement of an element's ability to conduct electric current. The unit of conductance is the Siemens (S).

2.3   Kirchhoff's Law

Network variables may have many interrelationships among themselves. Some of these relationships originate from the nature of the variables. A different class of relationship occurs because of the restriction that some specific type of network element places on the variables. Another class of relationship is one between several variables of the same type which occurs as the result of the network configuration, i. e. , the manner in which the various element of the network are interconnected. Such a relation is said to be based on the topology of the network. Kirchhoff's Current and Voltage Laws are laws based on the connective features of a network.

2.3.1   Kirchhoff's Current Law

Kirchhoff's Current Law is based on the law of conservation of charge, which requires an invaviant algebraic sum of charges within a system.

Kirchhoff's Current Law (KCL) states that the algebraic sum of currents entering a node (or a closed boundary) is zero. Mathematically, KCL implies that

$$\sum_{n=1}^{N} i_n = 0 \tag{2.5}$$

where N is the number of branches connected to the node and i_n is the n_{th} current entering (or leaving) the node. By this law, polarity of currents entering a node is regarded as " + ", while polarity currents leaving the node is taken as " − ".

On the consideration of the node in Figure 2.11, application of KCL gives

$$i_1 + (-i_2) + i_3 + i_4 + (-i_5) = 0 \tag{2.6}$$

Since current i_1, i_3 and i_4 are entering the node, while currents i_2 and i_5 are leaving it. By rearranging the Eq. (2.6), it can be obtained that:

$$i_1 + i_3 + i_4 = i_2 + i_5 \tag{2.7}$$

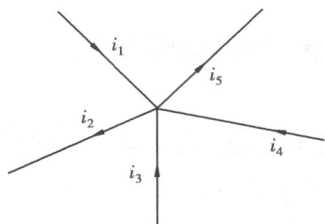

Figure 2.11   Current at a node illustrating KCL   Figure 2.12   Applying KCL to a closed boundary

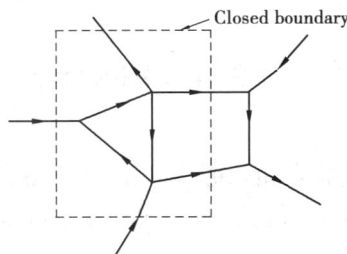

An alternative form of KCL: The sum of the currents entering a node is equal to the sum of the currents leaving it.

KCL also applies to a closed boundary. This may be regarded as a generalized case, because a node may be regarded as a closed surface shrunk to a point. In two dimensions, a closed boundary is the same as a closed path. As typically illustrated in the circuit of Figure 2.12, the total current entering the closed surface is equal to the total current leaving the surface.

2.3.2 Kirchhoff's Voltage Law

Kirchhoff's Voltage Law is based on the principle of conservation of energy.

Kirchhoff's Voltage Law (KVL) states that the algebraic sum of all voltages around a closed path (or loop) is zero. Mathematically, KVL states that

$$\sum_{m=1}^{M} u_m = 0 \qquad (2.8)$$

where M is the number of voltages in the loop and u_m is the m_{th} voltage.

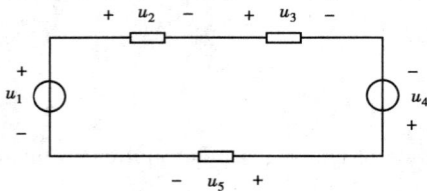

The circuit in Figure 2.13 illustrates KVL. The sign on each voltage is the polarity of the terminal encountered first as it travel around the loop. Starting from any node, turn the loop clockwise or counter clockwise. Suppose start with a voltage source and rotate clockwise around the loop as shown in the figure, then the voltages will be $-u_1$, $+u_2$, $+u_3$, $-u_4$ and $+u_5$ in order.

Figure 2.13　A single-loop circuit illustrating KVL

For example, when element 3 is reached, the positive end is first satisfied; therefore, there is $+u_3$. For element 4, the negative end is reached first; therefore, there is $-u_4$. Thus, KVL yields

$$-u_1 + u_2 + u_3 - u_4 + u_5 = 0 \qquad (2.9)$$

Rearranging terms gives

$$u_2 + u_3 + u_5 = u_1 + u_4 \qquad (2.10)$$

which may be interpreted as

Sum of voltage drops = Sum of voltage rises

This is an alternative form of KVL. If travel counterclockwise, the result would be u_1, $-u_5$, $+u_4$, $-u_3$, and $-u_2$, which is the same as before except that the signs are reversed. Hence, Eq. (2.9) and Eq. (2.10) are the same.

2.4　Electric power and sign convention

It is defined that the electrical power dissipated or stored by a passive element is equal to the product of the voltage across the element and the current flowing through it.

$$P = UI \qquad (2.11)$$

The unit of voltage (joules/coulomb) times current (coulombs/second) are unit of power (joules/second, or watts).

Just like voltage, power is signed quantity. It is necessary to make a distinction between positive and negative power. This distinction is shown in Figure 2.14, in which a source and a load are shown side by side. The polarity of the voltage across the source and the direction of the current through it indicate that the voltage source is driving charges to more from a lower potential to a higher potential. On the other hand, the load is dissipating energy, because the direction of the current indicates that charge is being displaced from a higher potential to a lower potential. To avoid confusion with regard to the sign of power, the electrical engineering community uniformly adopts the passive sign convention, which simply states that the power dissipated by a load is positive quantity

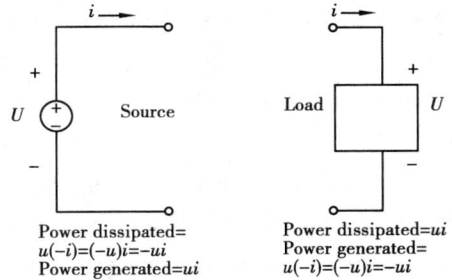

Power dissipated=
$u(-i)=(-u)i=-ui$
Power generated=ui

Power dissipated=ui
Power generated=
$u(-i)=(-u)i=-ui$

Figure 2.14 the passive sign convention

(or, conversely, that the power generated by a source is negative quantity). Another way of phrasing the same concept is to state that if current flows from a higher to a lower voltage, the power will be dissipated to be positive quantity.

The actual numerical values of voltages and currents are not very significant or easy to settle down at first, since once the proper reference directions have been established and the passive sign convention has been applied consistently, the answer will be correct regardless of the reference direction chosen.

2.5 Basic analysis methods

The laws of circuit theory (Ohm's Law and Kirchhoff's Laws) will be applied to develop two powerful solving techniques for circuit analysis: nodal analysis which is on the basis of a systematic application of Kirchhoff's Current Law (KCL), and mesh analysis based on a systematic application of Kirchhoff's Voltage Law (KVL). With the two techniques in this section, almost any circuit by obtaining a set of simultaneous equations solved to determine the required values of current or voltage can be analyzed. One method of solving simultaneous equations involves Cramer's Rule, which requires calculating circuit variables as quotient of determinants.

2.5.1 Nodal analysis

A convenient choice of voltages for many networks is the set of node voltages. Since a voltage is defined as potential difference between two nodes, it is convenient to select one node in the network as a reference node or datum node and then associate a voltage or a potential with each of the other nodes. The voltage of each of the non-reference nodes with respect to the reference node is defined to be a node voltage. It is common to select polarities so that the node voltages are positive relative to the reference node. For a circuit containing N nodes, there will be $N-1$ node voltages, some of which may be known, of course, if voltage sources are present.

Commonly the reference node is chosen to be the node to which the largest number of branches are connected. Many practical circuits are built on a metallic base or chassis, and usually there are a number of elements connected to the chassis, which is often then connected to the earth. The chassis may then be called ground, and it becomes the logical choice for the reference node. For this reason, the reference node is frequently referred to as ground. The reference node is thus at ground potential or zero potential, and the other nodes may be considered to be at some potential above zero.

The application of KCL results in an equation relating node voltages. Clearly, simplification in writing the resulting equations is possible when the reference node is chosen to be a node with a large number of elements connected to it. However, this is not the only criterion for selecting the reference node, it is frequently the overriding one.

In the network shown in Figure 2.15, there are three nodes. Since there are four branches connected to node 3, node 3 is selected as reference node to be identified by the sign of ground connection.

The voltage between node 1 and the reference node 3 is identified as u_1, and u_2 is defined between node 2 and the reference node 3. Notating two voltages are enough, because and the voltage between any other pair of nodes may be found in terms of them. For example, the voltage between node 1 and node 2 is $(u_1 - u_2)$.

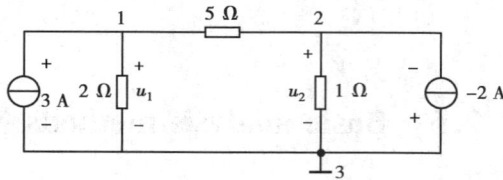

Figure 2.15   A given three-node circuit

Kirchhoff's Current Law is applied to node 1 and node 2. It can be implemented by equating the total current that leaving the node through the several conductances to the total source current entering the node. Thus,

$$0.5u_1 + 0.2(u_1 - u_2) = 3 \qquad (2.12)$$

or

$$0.7u_1 - 0.2u_2 = 3 \qquad (2.13)$$

At node 2, according to KCL, it can be obtained that

$$u_2 + 0.2(u_2 - u_1) = 2 \qquad (2.14)$$

or

$$-0.2u_1 + 1.2u_2 = 2 \qquad (2.15)$$

Solve the Eq. (2.13) and Eq. (2.15) to obtain the unknown node voltage u_1 and u_2, then any current or power in the circuit may be found out.

Steps to nodal analysis:

(1) Select a node as the reference node. Assign voltages u_1, u_2, $\cdots$, u_{n-1} to the remaining $n-1$ node.

(2) Apply KCL to each of the $n-1$ non-reference nodes. Use Ohm's Law to express the branch currents in terms of node voltages.

(3) Solve the resulting simultaneous equations to obtain the unknown node voltages, then solve

20

the other required variables.

2.5.2　Mesh analysis

Mesh analysis provides another general procedure for analyzing circuits, using mesh currents as the circuit variables. Using mesh currents, instead of 'element currents as circuit variables, is convenient to reduce the number of equations that must be solved simultaneously. A loop is a closed path with no node passed more than once. A mesh is an independent loop that does not contain any other loop within i_t.

Nodal analysis applies KCL to find unknown voltages in a given circuit, while mesh analysis applies KVL to find unknown currents. Mesh analysis is not quite as common as nodal analysis because it is only applicable to a circuit that is planar. A planar circuit is one that can be drawn in a plane with no branches crossing one another; otherwise it is nonplanar. A circuit may have crossing branches and still be planar if it can be redrawn with no crossing branches.

In Figure 2.16, for example, there are two meshes in this circuit. The current through a mesh is known as mesh currents in a given circuit. If the left-hand mesh is labelled as mesh 1, then a mesh current i_1 flowing in a clockwise direction along this mesh will be specified.

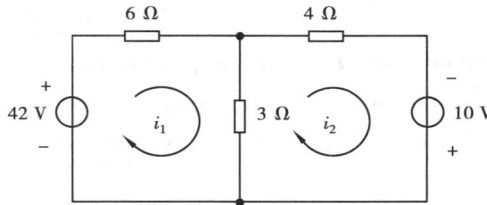

Figure 2.16   A mesh current in circuit

A mesh current is indicated by a curved arrow that almost closes on itself and is drawn inside the appropriate mesh, as shown in Figure 2.16. The mesh current i_2 is established in the remaining mesh, again in a clockwise direction. Although the direction is arbitrary, clockwise mesh currents are always chosen since a certain error-minimizing symmetry results in the equations.

One of the greatest advantages in the use of mesh currents is the fact that Kirchhoff's Current Law is automatically followed. If a mesh current flows into a given node, it obviously flows out of the node also.

Applying KVL to each mesh, it can be obtained that:

$$-42 + 6i_1 + 3(i_1 - i_2) = 0 \qquad (2.16)$$

$$\text{or}$$

$$9i_1 - 3i_2 = 42 \qquad (2.17)$$

$$-3i_1 + 7i_2 = 10 \qquad (2.18)$$

In Eq. (2.17), the coefficient of i_1 is the sum of the resistances in the mesh 1, while the coefficient i_2 is the negative polarity of the resistance in both meshes 1 and 2. It is the same in Eq. (2.18).

The branch currents are different from the mesh currents unless the mesh is isolated.

Steps to mesh analysis:

(1)Assign mesh current i_1, i_2, ... , into the n meshes respectively.

(2)Apply KVL to each of the n meshes. Use Ohm's Law to express the voltages in terms of the mesh currents.

(3)Solve the resulting n simultaneous equations to the mesh currents, and then solve the other required variables.

2.5.3 Superposition principle

Superposition principle is frequently applied for the analysis of linear circuits. Rather than a precise analysis technique, like the mesh current and node voltage methods, the principle of super-position is a conceptual aid that can be very useful in visualizing the behavior of a circuit containing multiple sources. It is applied to any linear system and for a linear circuit. The solving technique may be stated as follows: In a linear circuit containing N sources, each component voltage and branch current is the sum of N parts of voltages and currents, each parameter may be computed by setting all but one source to zero and solving the circuit containing that single source.

It states that the total voltage across (or current through) an element in a linear circuit is the algebraic sum of the voltage across (or currents through) that element due to each independent source acting alone.

The principle of superposition helps us to analyze a linear circuit with more than one independent source by calculating the contribution of each independent source separately.

Consider the effects of 8 A and 20 V individualy, then add the two response together to determine the total value of U_0.

For example, in Figure 2.17, the circuit with two sources, U_0 must be determined.

The solving technique is to analyze the response U_0 of two sources namely 8 A or 20 V.

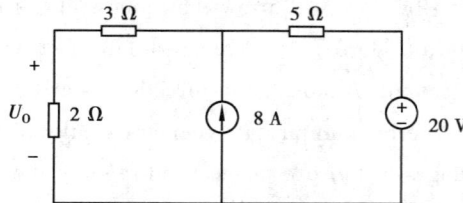

Figure 2.17   A given circuit with two sources

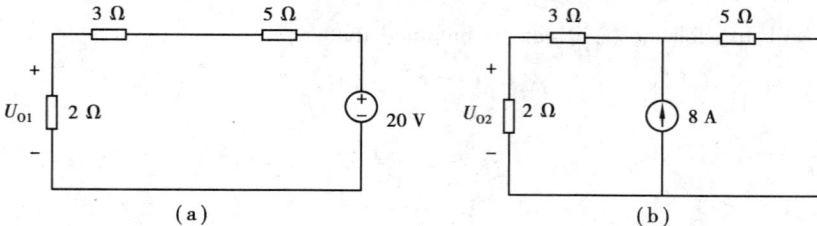

(a)   (b)

Figure 2.18   aportion of circuit with a single source

According to Figure 2.18, We get

$$U_{01} = 4 \text{ V} \tag{2.19}$$

$$U_{O2} = 8 \text{ V} \qquad\qquad (2.20)$$
$$U_O = U_{O1} + U_{O2} = 10 \text{ V} \qquad\qquad (2.21)$$

Steps to apply superposition principle:

(1) Draw a portion of circuit by zeroing, which means independent current source must be taken away, and voltage source is bypassed or short.

Three things have to be kept in mind when zeroing all other independent sources:

a. Independent voltage sources are replaced by 0 V (short circuit);

b. Independent current sources are replaced by 0 A (open circuit);

c. Dependent sources are left intact because they are controlled by circuit variables.

(2) Find the portion of output (voltage or current) by nodal or mesh analysis method.

(3) Repeat step 1 to draw circuit with different each part of independent sources.

(4) Find the total contribution by adding algebraically all the responses from each independent source.

2.5.4　Thévenin and Norton equivalent circuits

It states that a linear two-terminal circuit, shown as Figure 2.19 a, can be replaced by an equivalent circuit as Figure 2.19 b, which consists of a voltage source V_{TH} in series with a resistor R_{TH}, where V_{TH} is the open-circuit voltage between two terminals, R_{TH} is the input or equivalent resistance at the terminals when all the independent sources inside the black box are zeroed.

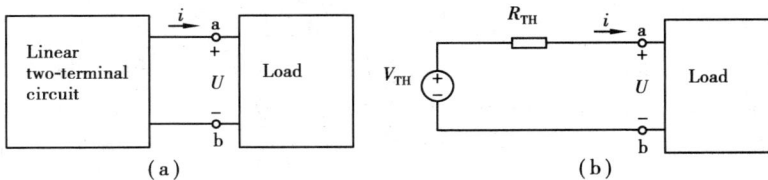

Figure 2.19　Thévenin equivalent circuit of A linear two-terminal circuit

2.5.5　Norton's theorem

It states that a linear two-terminal circuit can be replaced by an equivalent circuit of a current source I_N in parallel with a resistor R_N, where I_N is the short circuit current through the terminals, R_N is the input or equivalent resistance at the terminals when the independent sources inside the black box are zeroed, as shown in Figure 2.20.

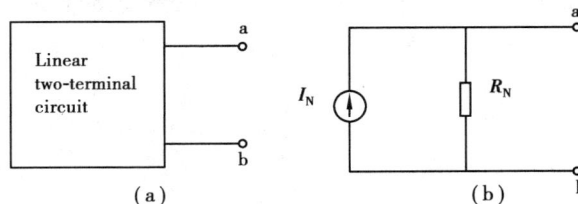

Figure 2.20　Norton's Theorem

2.6　Sinusoidal AC circuit and three-phase circuit

In a sinusoidally excited linear circuit, all element voltages and branch currents are sinusoids at the same frequency as the excitation signal. The amplitudes of these voltages and currents are scaled versions of the excitation amplitude, and the voltages and currents may be shifted in phase with respect to the excitation signal.

These observations indicate that three parameters uniquely define as sinusoid: frequency, amplitude, and phase. But if this is the case, is it necessary to carry the "excess luggage", that is, the sinusoidal functions? Might it be possible to simply keep track of the three parameters just mentioned? The answers to these two questions are no and yes, respectively. The next section will describe the use of a notation that, with the aid of complex algebra, eliminates the need for the sinusoidal functions of time, the formulation and solution of differential equations, permitting the use of simpler algebraic methods.

Any sinusoidal signal may be mathematically represented in two ways: one is time-domain form,

$$u(t) = A \cos(\omega t + \varphi) \tag{2.22}$$

and the other is frequency-domain (or phasor) form,

$$U(j\omega) = A e^{j\varphi} = A\angle\phi \tag{2.23}$$

Note the $j\omega$ in the notation $U(j\omega)$, indicating the $e^{j\omega t}$ dependence of the phasor. In the remainder of this chapter, bold uppercase quantities will be employed to indicate phasor voltages or currents.

A phasor is a complex number, expressed in polar form, consisting of a magnitude equal to the peak value of the sinusoidal signal and a phase angle equal to the phase shift of the sinusoidal signal referenced to a cosine signal.

When using phasor notation, it is important to make a note of the specific frequency, ω, of the sinusoidal signal, since this is not explicitly apparent in the phasor expression.

2.6.1　Phasor relationships for circuit elements

Simplification of sinusoidal steady-state analysis can be proceeded by establishing the relationship between the phasor voltage and phasor current for each of the three passive load.

The resistor provides the simplest case. In the time domain, as indicated by Figure 2.21(a), if the current through a resistor R is a sinusoidal signal, the voltage across R is given by Ohm's law as

$$u = Ri(t) = RI_\mathrm{m} \cos(\omega t + \varphi) \tag{2.24}$$

The phasor form of this voltage is

$$\dot{U}_\mathrm{m} = RI_\mathrm{m}\angle\varphi = R\,\dot{I}_\mathrm{m} \tag{2.25}$$

$$\dot{U} = R\dot{I} \qquad\qquad (2.26)$$

In Figure 2.21(b), the voltage-current relation for the resistor in the phasor domain still obey Ohm' law, as in the time domain. As shown in Eq. (2.25), the phasor voltage, and current $\dot{I}$ are in phase, as illustrated in the phasor diagram in Figure 2.22.

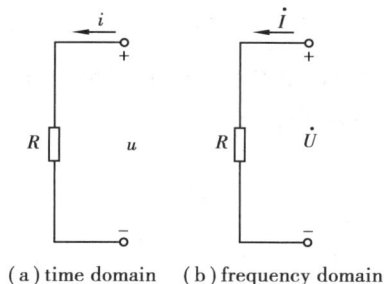

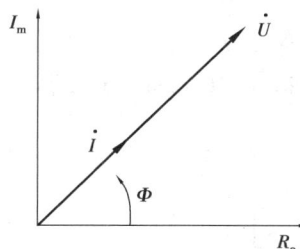

(a) time domain   (b) frequency domain

Figure 2.21   Voltage-current relation for the resistor

Figure 2.22   Phasor diagram for the resistor

For the inductor L, assuming the current through it is

$$i = I_{\mathrm{m}} \cos(\omega t + \varphi) \qquad\qquad (2.27)$$

The voltage across the inductor is

$$u = L\frac{\mathrm{d}i}{\mathrm{d}t} = -\omega L I_{\mathrm{m}} \sin(\omega t + \varphi) \qquad\qquad (2.28)$$

The voltage is re-written as

$$u = \omega L I_{\mathrm{m}} \cos(\omega t + \varphi + 90°) \qquad\qquad (2.29)$$

Then the phasor from is

$$\dot{U} = \omega L I_{\mathrm{m}} \mathrm{e}^{j(\varphi+90°)} = \omega L I_{\mathrm{m}} \mathrm{e}^{j\varphi} \mathrm{e}^{j90°} = \omega L I_{\mathrm{m}} \angle \varphi \ \mathrm{e}^{j90°} \qquad\qquad (2.30)$$

But $I_{\mathrm{m}} \angle \phi = \dot{I}$, $\mathrm{e}^{j90°} = j$,

Thus,

$$\dot{U} = j\omega L \dot{I} \qquad\qquad (2.31)$$

It shows that the voltage has a magnitude of $\omega L I_{\mathrm{m}}$ and a initial phase of $+90°$. The voltage leads current by $90°$. Figure 2.23 shows the voltage-current relations for the inductor. Figure 2.24 shows the phasor diagram.

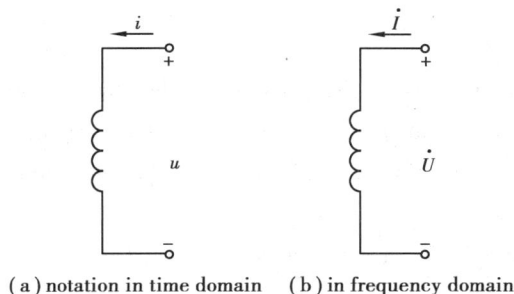

(a) notation in time domain   (b) in frequency domain

Figure 2.23   Voltage-current relation for an inductor

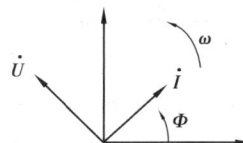

Figure 2.24   Phasor diagram for the inductor

For the capacitor C, assume the voltage across it is $u = U_m \cos(\omega t + \Phi)$. Then, current through the capacitor is

$$i = C \frac{du}{dt} \tag{2.32}$$

By following the same steps as for the inductor, it can be obtained that

$$\dot{I} = j\omega C \dot{U} \tag{2.33}$$

To be specific, the current leads the voltage by 90°. Figure 2.25 shows notation of the voltage-current in two kinds of domain, Figure 2.26 shows the phasor diagram.

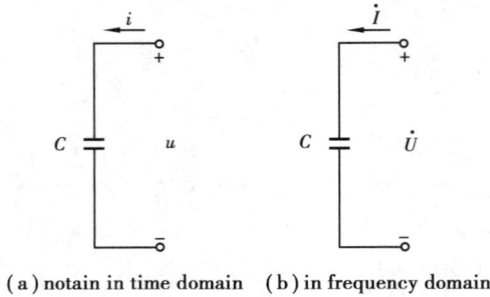

(a) notain in time domain (b) in frequency domain

Figure 2.25   voltage-current relation
for the capacitor

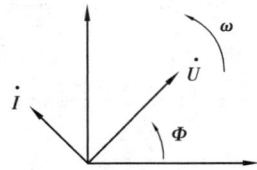

Figure 2.26   phasor diagram
for the capacitor

2.6.2　Analysis procedures for sinusoidal AC circuit

Ohm's Law and Kirchhoff's Law are applicable to AC circuits. The methods of simplifying circuit analysis (such as nodal analysis, mesh analysis, Thévenin's theorem and so on) are still applied in analyzing AC circuits.

There are three steps in analyzing AC circuit.

(1) Transform the circuit variables from time domain to the phasor or frequency domain.

(2) Solve the problem by using circuits solving techniques (nodal analysis, mesh analysis, superposition theorem etc.).

(3) Transform the response in phasor back to the time domain, if it is necessary.

2.6.3　Balanced three-phase voltages

A typical three-phase system consists of three voltage sources connected to loads by three or four wires (or transmission lines). A three-phase system is equivalent to three single-phase circuits. The voltage sources can be either wye-connected as shown in Figure 2.27 (a) or delta-connected as in Figure 2.27 (b).

Let's take the wye-connected voltages in Figure 2.27 (a) into consideration. The phase voltages $\dot{U}_{an}$, $\dot{U}_{bn}$ and $\dot{U}_{cn}$ are phasor voltage difference respectively between line a, line b, line c and the neutral lines n. These voltages are called phase voltages. If these three voltage sources have the same amplitude and frequency ω, and the phase difference is 120°, then we call it voltages balance. This implies that

$$\dot{U}_{an} + \dot{U}_{bn} + \dot{U}_{cn} = 0 \tag{2.34}$$

$$|\dot{U}_{an}| = |\dot{U}_{bn}| = |\dot{U}_{cn}| \tag{2.35}$$

Since the three-phase voltages are 120° out of phase with each other, there are two possible combinations. One possibility is shown in Figure 2.28 (a) and expressed mathematically as

$$\dot{U}_{an} = U_p \angle 0° \tag{2.36}$$

$$\dot{U}_{bn} = U_p \angle -120° \tag{2.37}$$

$$\dot{U}_{cn} = U_p \angle -240° = U_p \angle 120° \tag{2.38}$$

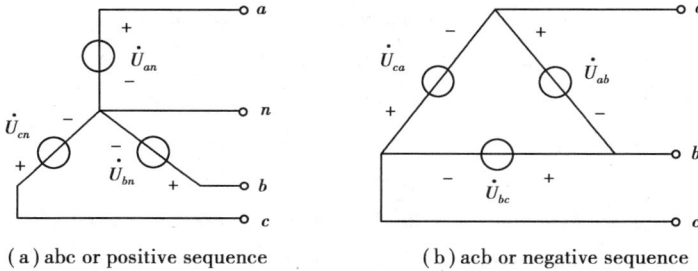

(a) abc or positive sequence   (b) acb or negative sequence

Figure 2.27   Phase sequences

This is known as the abc sequence or positive sequence. In this phase sequence, a leads b, and then b leads c. The other possibility is shown in Figure 2.27(b). This is called the acb sequence or negative sequence. For this phase sequence, a leads c, and then c leads b. The phase sequence is the time order in which the voltages pass through their respective maximum values. The phase sequence is determined by the order in which the phasors pass through a fixed point in the phasor diagram.

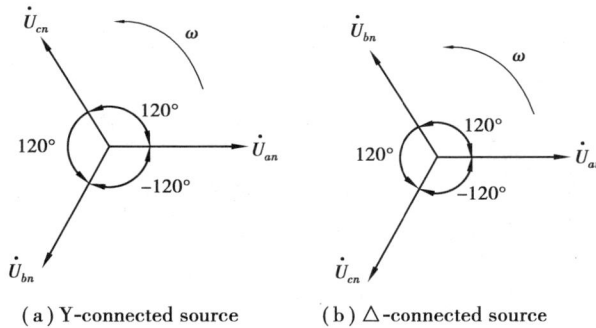

(a) Y-connected source   (b) $\triangle$-connected source

Figure 2.28   Three-phase voltage sources

The phase sequence which determines the direction of the rotation of a motor connected to the power source is important in three-phase power distribution.

Like the generator connections, a three-phase load can be either wye-connected or delta-connected, depending on the end application. Figure 2.29(a) shows a wye-connected load, and Figure

2.29(b) shows a delta-connected load. The neutral line in Figure 2.29(a) may not be there, depending on whether the system is four-or three-wire. (a neutral connection is topologically impossible for a delta connection.) A wye-connected or delta-connected load is unbalanced if the impedances are not equal in magnitude or phase angle. A balanced load is one in which the phase impedance is equal in magnitude and in phase angle.

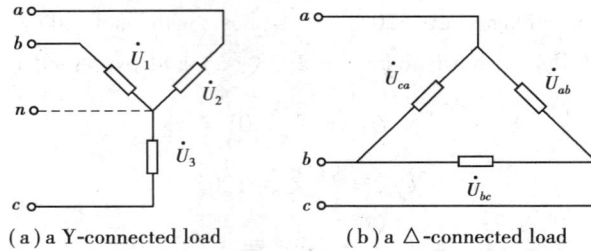

(a) a Y-connected load   (b) a △-connected load

Figure 2.29   Two possible three-phase load configurations

Since both the three-phase source and the three-phase load can be either wye-or delta-connected, there are four possible connections: Y-Y connection (i. e. Y-connected source with a Y-connected load); Y-△ connection; △-△ connection; △-Y connection.

A balanced delta-connected load is more common than a balanced wye-connected load. This is due to the ease with which loads may be added or removed from each phase of a delta-connected load. It is very difficult with a wye-connected load because the neutral may not be accessible. On the other hand, delta-connected sources are not common in practice because of the circulating current that will generate circulating current in the triangle connected mesh if the three-phase voltage are slightly unbalanced.

Chapter 3

Electric Machines and Apparatus

Electric motor is a kind of electromagnetic device which transforms or transfers electric energy according to the law of electromagnetic induction and the law of electromagnetic force. When the electric machine is used for energy conversion, it should have two components that can realize relative motion, that is, the components of the excitation magnetic field and the induction parts that generate electromotive force and flow through the working current. Of these two components, the static winding is called the stator, and the rotating winding is called the rotor (the one side of the straight line motion is also called the rotor). There is an air gap between the stator and the rotor to make it possible for the rotor to rotate (or reciprocate).

The electromagnetic torque is generated by the interaction between the excitation magnetic field in the air gap and the magnetic field created by the current in the induction component. Through the action of electromagnetic torque, the generator absorbs mechanical power from the mechanical system, and the motor outputs mechanical power to the mechanical system. The two magnetic fields are created in different ways to form different kinds of motors. If two magnetic fields are both generated by DC current, a DC motor is formed; if two magnetic fields are generated by alternating current of different frequencies, an asynchronous motor is formed; if one magnetic field is generated by both DC current and alternating current, a synchronous motor is formed.

When the current flows through the winding, it will produce a certain flux linkage, and its coupled magnetic field will store certain amount of electromagnetic energy. The flux and magnetic field energy varies with the stator, rotor current and rotor position. Electromotive force and electromagnetic torque are generated so as to realize electromechanical energy conversion. This energy conversion is reversible, that is, the same electric machine can be used either as a generator or as an motor.

3.1　Classification of electric machine

Electric machine is the general name of electromagnetic mechanical equipment for electromechanical energy conversion or signal conversion. According to different angles, there are different classification methods:

According to the type of current used, it can be divided into DC motor and AC electric machine.

According to the function in the application, it can be divided into the following categories:

①A generator that converts mechanical energy into electrical energy.

②A motor that converts electrical energy into mechanical energy.

③There are four kinds of power converters for converting electric energy into another one: Ⅰ transformers with different output and input voltages. Ⅱ A converter whose output and input waveforms are different, such as converting alternating current to direct current. Ⅲ Frequency converter with different output and input frequency. Ⅳ Phase converter with a different phase between the output and the input.

④A control motor that regulates, amplifies, and controls in an electromechanical system.

According to the running speed, motors can be divided into five types:

①Static equipment-transformer.

②There is no fixed synchronous speed-DC motor.

③Rotor speed is always different from synchronous speed asynchronous motor.

④Speed is equal to synchronous speed-synchronous motor.

⑤Wide speed range-AC commutator motor.

According to the size of power, it can be divided into large motor, small motor and micro motor.

With the development of power electronic materials, there are some special motors which do not belong to the traditional motor types. These motors are usually called special motors, including stepper motors, brushless motors, switched reluctance motors and ultrasonic motors.

3.2　Generator

Generator is widely used in industrial and agricultural production, national defense, science, technology and daily life. There are many forms of generators, and their working principles are based on electromagnetic induction law and electromagnetic force law. Therefore, its structure follows general principles. Suitable magnetic and conductive materials constitute magnetic circuit. They can be electro magnetically induced by each other. Therefore, electromagnetic power can be produced to achieve the purpose of energy conversion.

The generator is usually composed of stator, rotor, end cover and bearing. The stator is composed of a stator core, a wire winding, a frame, and other structures that fix these parts. The rotor is composed of a rotor core (or magnetic pole, magnetic choke), winding, retaining ring, central ring, slip ring, fan, a rotating shaft and other components.

The stator and rotor of the generator are connected by the bearing and the end cover, so that the rotor can rotate in the stator, cutting magnetic induction line, and the induced electromotive force is generated. So the current is produced in the loop.

The generator can be divided into DC generator and AC generator. AC generator includes synchronous generator and asynchronous generator, AC generator can also be divided into single-phase generator and three-phase generator. These generators, which must rotate at a fixed angular speed and produce the same frequency of alternating current at any time, are connected to the AC network. This kind of motor is called synchronous generator.

Large generators are mainly synchronous generators with a single capacity of one hundred thousand kilowatts at most. Small capacity generators are used in stand-alone power systems, such as diesel generators, wind turbines, etc. Because the synchronous generator needs the excitation set, in some occasions, the wind power generator also can use the asynchronous generator to generate electricity. Modern power plants no longer use DC generators, only small DC generators are used on some special occasions.

The stator of large synchronous generator is made up of silicon steel sheet and the symmetrical three-phase winding which is placed in the groove of the iron core. The rotor is made of ferromagnetic material, and the excitation winding is placed in it. The excitation winding is connected to the DC excitation power supply through the slip ring. When the generator is driven by the prime mover, the rotating fields cuts the conductor of the three-phase winding and generates the induced electromotive force in the winding.

Because the stator winding is a symmetrical three-phase winding, the electromotive force induced by the stator is symmetrical three-phase potential. There are two types of rotors for large synchronous generators: the implicit pole type and the salient pole type. The implicit pole rotor is cylindrical, and the air gap of the generator is uniform air gap. This kind of generator is mostly used in high speed and large capacity turbogenerator. The salient pole rotor is mostly used for hydro generator. The speed of the generator is lower, and the number of poles of the motor is bigger. The rotor is connected with the prime mover through the shaft.

3.2.1　Turbogenerator

Turbogenerator is a generator driven by steam turbine. Superheated steam from the boiler enters the turbine to expands and rotate the blades, and to drive the generator. The exhaust air is turned to the boiler for circulation through the condensers, circulating pumps, condensate pumps, water supply heating devices, etc.

The generator in thermal power plants and nuclear power plants is a kind of synchronous generator. The prime movers (turbine) rotate at very high speed, usually 3 000 revolutions per minute (at a frequency of 50 Hz). Therefore, in order to reduce the mechanical stress caused by centrifugal force and reduce the wind abrasion, the rotor diameter of the turbogenerator is generally small and its length is large (i. e. the slender rotor), so as to reduce the linear speed. After 1970s, the maximum output power of the turbinegenerator reached 130-150 kilowatts. Steam turbine generator in thermal power plant as shown in Figure 3.1.

(a) Thermal power plant

(b) Steam turbine generator

(c) Stator of turbogenerator

(d) Rotor of turbogenerator

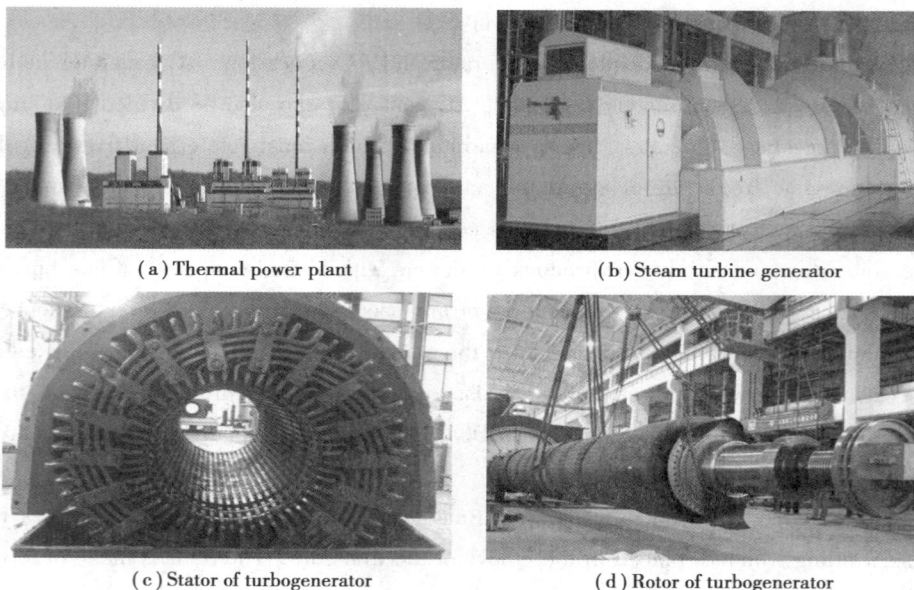

Figure 3.1　Steam turbine generator in thermal power plant

3.2.2　Hydrogenerator

A hydrogenerator is a generator that can convert water energy into electricity through a hydroturbine. When water passes through the turbine, the water energy is converted into mechanical energy. The mechanical energy can be converted into electrical energy through the rotor drived by rotating shaft of the hydroturbine.

Hydrogenerator is the main power equipment to produce electric energy in hydraulic power plants. Hydrogenerator is also a kind of synchronous generator. Because the prime mover (turbine) has a lower speed, generally less than a few hundred RPM, the rotor of the generator is thick and short to increase the number of poles. The starting and grid connection time is short, so the operation scheduling is flexible. Besides the general power generation, the hydrogenerator unit is especially suitable for peaking unit and emergency standby unit. The maximum power of a single hydrogenerator unit has reached 800 thousand kW. The total installed hydrogenerator capacity of the Three Gorges power station has 26 units, with a capacity of 700 thousand kW per unity, as shown in Figure 3.2.

Figure 3.2　Hydrogenerator of Three Gorges power plant

3.2.3   **Wind power generator**

The wind turns the wind energy into mechanical energy through the impeller device, and then the mechanical energy can been turned into electrical energy. Electricity is generated in this process.

Wind energy is one of the cheapest, cleanest and most valuable energy sources. In recent decades, many countries have started to accelerate the development of wind power. With the development of large wind turbines, many wind farms have been built, and the wind power industry has grown rapidly. The wind turbine generator as shown in Figure3.3.

Figure 3.3   Wind turbine generator

3.3   **Transformer**

3.3.1   **Structure and function**

Transformer is a device to change the amplitude of AC voltage according to the principle of electromagnetic induction. It is a static electric machine. The transformer consists of an iron core (or magnetic core) and a coil with two or more windings. The winding of the power supply is called a primary coil, the rest of the winding called a secondary coil.

The simplest core transformer consists of an iron core which is made of a soft magnetic material and two coils with different turns harnessed to the iron core. The role of the core is to strengthen magnetic coupling among the two coils. In order to reduce the eddy current and the hysteresis loss in the iron, the iron core is made up of the silicon steel sheet coated with insulating paint, and there is no electric connection between the two coils. The coil is made of insulated copper wire (or aluminum wire). In order to improve the heat dissipation condition, the core and winding of large and medium capacity power transformers are immersed in the closed oil tank filled with oil, and the electric connection between the internal and external windings is led by the insulating bushing. In order to ensure the safe and reliable operation of the transformer, there are also oil conservator, safety airway, gas relay and other accessories. Figure 3.4 shows the three-phase oil immersed power transformer and its internal structure.

Figure 3.4   Three-phase oil immersed power attain transformer

The transformer can change the magnitude of AC voltage and current besides impedance trans-formation, it may isolate and stabilize the voltage(namely magnetic saturation transformer). The transformer can only transfer AC energy, but can't generate electricity. It cannot change the frequency. Transformer is one of the most important equipments, used in substation and power distribution in power system, as shown in Figure 3.5.

Figure 3.5   Transformer in substation

3.3.2   Classification

There are many kinds of transformers, which can be classified according to phase number, us-age, winding form, structure of iron core, cooling mode, etc.

(1) **According to the phase number**

① Single phase transformer: for single phase load and three-phase transformer set.

② Three-phase transformer: three-phase system for the rise and fall voltage.

(2) **According to the usage**

① Power transformer: it can be used in the power transmission and distribution system for the rising and falling of voltage. Because of the most widely usage, its production is the largest. Accord-ing to its function, it can be divided into step-up transformer, step-down transformer, distribution transformer and so on. The apparent power of power transformer ranges from tens of VA to dozens of kVA, while voltage is from several hundred Volt to a thousand kV.

② Instrument transformer: such as voltage transformer, current transformer, they are used for measuring instruments and relay protection devices.

③ Test transformer: it can be produced to output relatively high voltage, so it is an important component of the electrical equipment for high voltage test.

④ Special transformer: it can been used on special occasions, including welding power transformer, furnace transformer for high power electric furnace, rectifier transformer which can convert AC into DC, shifting transformer and so on.

(3) According to the winding form

① Double winding transformer: it connects two voltage levels in power systems.

② Three winding transformer: it is commonly used in regional substation of power system for connecting three voltage levels.

③ Autotransformer: there are many kinds of transformers in power systems for connecting different voltages. It can also be used as ordinary step-up or stop down transformer.

(4) According to the structure of iron core

① Core transformer: it can be used in high voltage substation.

② Amorphous alloy transformer: an Amorphous core transformer is a new magnetic material with the no-load current under 80%. Its distribution transformer has an ideal energy saving effect, especially suitable for places with low load rates in rural and less developed areas.

③ Shell transformer: a special transformer for high current, such as furnace transformer, welding transformer, power transformer for electronic instruments, television, radio and so on.

(5) According to cooling mode

① Dry-type transformer: it relies on air convection for natural cooling or fan cooling, often used to supply enengy for high-rise buildings, high speed toll station power, local lighting, electronic circuits and other small capacity transformers.

② Oil-immersed transformer: it depends on oil as cooling medium, such as oil-immersed self-cooling, oil immersed air cooling, oil immersed water cooling, forced oil circulation, etc.

3.4 Motor

3.4.1 Overview of motors

An electric motor is a machine that converts electrical energy into mechanical energy. It uses the energized coil (i. e. stator winding) to produce a rotating magnetic field and acts on the rotor to form Magnetoelectric rotating torque. As shown in Figure 3.6, the motor is mainly composed of stator and rotor, and the direction of the force acting on the conducting wire in the magnetic field is related to the direction of the current and the magnetic line of force (the direction of the magnetic field). The working principle of the motor is to make the motor rotate under the effect of the magnetic field on the current.

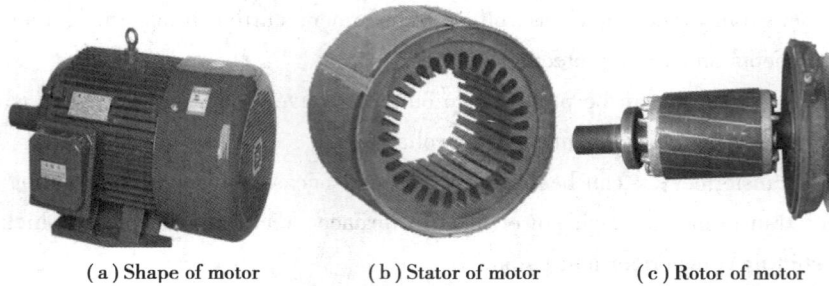

(a) Shape of motor   (b) Stator of motor   (c) Rotor of motor

Figure 3.6   Structure of motor

Motor drive is widely used in all kinds of modern production machinery. The production machinery driven by electric motor has many advantages, it can simplify the structure of production machinery, improve the productivity and quality of products, realize automatic control and remote control, and reduce the demand of labor force. Among them, the small power motors and micro motors are used for electric tools and household appliances, but also used in the automatic control system and a computing device as detection, amplification, actuator etc. Motors are divided into DC motor and AC motor according to the different power supply, and the latter one is common in power system.

DC motor has better characteristics than AC motor as wide speed range, stable speed regulation, high overload capacity, large starting and braking torque. It is widely used in the production machinery with high requirements for starting and speed regulation. DC motors are often used as traction motors and auxiliary compressors in domestic electric locomotives.

The main function of AC asynchronous motor is to drag all kinds of production machinery. Asynchronous motor includes three-phase asynchronous motor and single-phase asynchronous motor. Compared with other motors, it has a series of advantages, such as simple structure, easy to manufacture, low price, reliable operation, convenient maintenance and high efficiency. That is why asynchronous motor has been widely used. At present, the most popular motor used in production is three-phase induction motor, accounting for more than 60% motors in the world. The structure of the three-phase induction motor is shown in Figure 3.7.

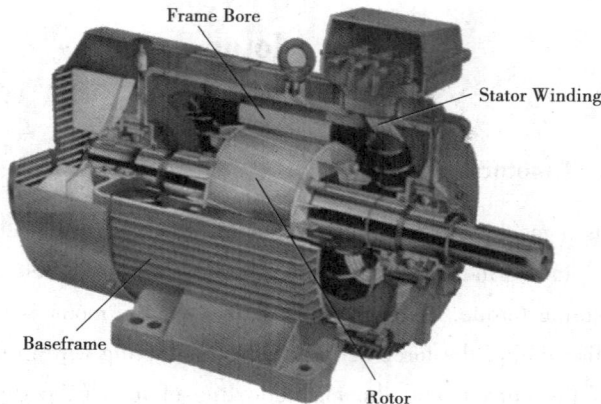

Figure 3.7   Structure of three phase-induction motor

3.4.2   The application of motor

The asynchronous motors used in industry are medium and small-sized rolling equipments as shown in Figure 3.8(a): They are used in various metal cutting machine tools, light industry machinery, mining machinery, etc.

The asynchronous motors are used in agriculture, including pumps as shown in Figure 3.8(b), threshing machines, shredders, and other agricultural and sideline products processing machinery. In addition, the asynchronous motor is used in the fans, washing machines and other househlold appliances. Building water supply, heating, ventilation and so on needs pumps and blowers as shown in Figure 3.8(c) that are also driven by asynchronous motor. In the electric locomotive, the locomotive in operation, transformer, current transformer, traction motor and brake resistance device emits a lot of heat, therefore, the cooling fan is forced to be installed; forced circulation of transformer oil pump is needed; brake, locomotive and train pantograph have to be provided with air compressor etc. The disadvantage of asynchronous motor is that it can't economically smooth speed regulation in a large range, and the reactive power must be absorbed from the power grid to reduce the power factor of the grid.

(a) Rolling mill   (b) Water pump   (c) Blower

Figure 3.8   Machinery for Application of Asynchronous Motor

A modern robot uses a lot of motors for motion control. Some robot joints are directly spherical motors as shown in Figure 3.9, which can be used to realize universal motions.

Figure 3.9   Spherical motor

In high-rise buildings, elevators and escalator are drived by motors. The automatic doors and revolving doors shown in Figure 3.10 are also pulled by motors.

Figure 3.10   Revolving door

3.4.3   The selection of motor

The types of motors are selected from the aspects of AC or DC, mechanical characteristics, speed and starting performance, maintenance and price. Because there are three-phase AC power supply in produdtion site, if there is no special requirements, AC motor should be used generally. As in the AC motors, the structure of three-phase squirrel-cage asynchronous motor is simple, durable, reliable, cheap and easy for maintenance; its main shortcomings are difficult to control, low power factor and poor start-up performance. When starting, braking and speed of machinery have no special requirements, the general should adopt three-phase cage induction motor, for example, small power pumps, ventilation machine, conveyor, conveyor belt, machine tool auxiliary motion mechanism, almost all are equipped with squirrel-cage motors. Some spindle motors are also used in small machine tools. The heavy load starting of machinery, such as certain crane, hoist, forging machine and heavy machine tool must meet the requirements of starting or increasing unreasonable power or small speed range of machinery, and the low speed running time is short, so it should adopt the wound rotor electric machine. Synchronous motor should be used when machines are large and continuous. When the machinery has special requirements for starting, speed regulation and braking, the type of motor and its speed regulating mode should be determined according to the technical and economic comparison. When the AC motor can't meet the mechanical requirements of the characteristics, it is advisable to use DC motor; DC motors may alse be used in emergency units that have to work after the AC power runs out. When the fan and pump machinery with variable load are used, the speed regulating device should be adopted and the corresponding type motor should be selected in a technically and economically reasonable situation.

3.5   Special motor

Special motors usually refer to the ones which have different structure, performance, application or principle from the conventional motors, and the volume and output power of the small motor or special precision motor, generally its outer diameter is not greater than 130 mm. Special motors has two function: one is for driving and the other one is for controling, the former is mainly used to drive a variety of institutions, instruments and household appliances; the latter is a general term for small power motor transmission and transformation in the automatic control system and the implementation of the control signal used as actuators or signal components. Special motors for control are both measuring elements and actuating elements. The measuring elements include rotating transformer, AC and DC tachometer generator, etc. The main components of the actuator include AC and DC servo motors, stepper motors, etc.

3.5.1   Servo motor

A servo motor is a kind of small motor in the automatic control system, as shown in Figure 3. 11. It is also named as the actuator, performing transition actions such as start, stop, forward and backward according to input signals, driving operation and mechanical load. The speed and steering of the servo motor vary with the magnitude and polarity of the control voltage, and the input voltage signal is transformed into the angular displacement or angular velocity output of the rotating shaft. It is often used as main transmission components with pneumatic systems, telemetry and remote control systems and all kinds of incremental motion control systems. It is widely used in industrial robots, machine tools, measuring instruments, office equipments, such as printers and plotters device.

Figure 3. 11   Servo motor

A servo system of NC (Numerical Control) machine tool is an automatic control system which takes the mechanical displacement of moving parts of machine tool as the direct control target, also known as the position servo system. It receives the step pulse from the interpolator and transforms it into the displacement of the machine worktable after the magnification. The servo system of high performance CNC (Computer numerical Control) machine tool also feedback the actual output position state by the detecting elements, and then the position controller is adopted for the position closed-loop control, as shown in Figure 3.12.

Figure 3.12   Application of servo motor in CNC machine tool

3.5.2 **Linear motor**

A linear motor can convert the electric energy directly into the mechanical energy by straight line movement, without any intermediate conversion mechanism, as shown in Figure 3.13. It not only eliminates the mechanical transmission between rotary motor and linear working body of the device, but also puts one side of the linear motor in place as part of the mechanical motion to make the whole device compact and reasonable, at high efficiency and low cost. Therefore, it has been widely used in many fields of industry.

Figure 3.13   linear motor

The high speed train driven by linear motor, that is, maglev train is one of the typical examples. It can reach more than 400 km per hour. The so-called maglev train is to use the magnetic suspension car body and the linear motor drive technology to make the train float and move on the track, as shown in Figure 3.14. Magnetic field are created by the electromagnets below the track and on the roller coaster. The two magnetic fields attract each other. Above the linear motor rail mobile magnetic traction led the rear of the train move along the track at a high speed, as shown in Figure 3.15.

The application of linear motor in urban rail transit and high-speed railway is shown in Figure 3.16 and Figure 3.17. In recent years, the application of linear motor in industrial machinery, elevators, carrier thransmitter, electromagnetic guns, missile launchers, electromagnetic propulsion submarine has been widely found.

Figure 3.14   maglev train in Shanghai

Figure 3.15   roller coaster

Figure 3.16   light rail train in the city

Figure 3.17   high speed inter city train

3.5.3   Stepper motor

A stepper motor converts the electrical pulse signal into corresponding angular displacement. When an electric pulse is applied to the stepper motor control winding, the shaft rotate a certain angle. The angular displacement and electric pulse is proportional to the number of electric pulses, and the speed is proportional to the pulse frequency. Every time a pulse signal is input, the stepper motor moves forward one step, which is also called the stepper motor, as shown in Figure 3.18.

Figure 3.18   stepper motors

A stepper motor is driving by frequency pulse signal source, pulse distributor and power pulse amplifier. The electronic circuit will be used to put DC component, multi-phase timing control current output into a motor of a stepping power supply. The driving power supply provides the impulse current to the motor winding, so the running performance of the stepper motor depends on the good coordination between the motor and the driving power supply.

With the development of microelectronics and computer technology, the demand of stepper motor grows day by day. The application of stepper motor is very extensive, such as mechanical processing, drawing machine, robot, computer peripheral equipments, automatic recording instruments etc. It is mainly used in difficult, fast, high precision occasions. Especially the development of power electronics and microelectronics technology opens up a broad prospect for the application of stepper motors. In digital control systems, stepper motors are often used as actuators.

Here is a simple example of the application of the stepper motor with a CNC machine tool, as shown in Figure 3.19. The CNC machine tool is the abbreviation of numerical control machine. It has the characteristics of generality, flexibility and high automation. It is mainly applicable to the production of parts with high precision and complex shape. Its working process is: first of all, it should be in accordance with the machining requirements and machining process, the machining program should be input to the computer, and the computer calculates data and executes instructions according to the program; then according to the results obtained in every direction, stepper motor sends a corresponding control pulse signal, and drives the working mechanism according to the processing requirements in order to complete a variety of actions, such as speed change, reverse, stop, and so on, so it can automatically process the required program.

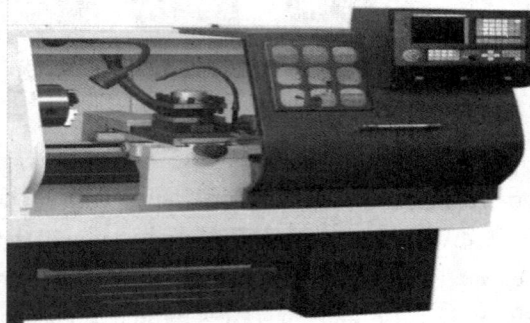

Figure 3.19　CNC machine tool

A stepper motor is a kind of induction motor, which has two basic types: electromechanical type and magnetoelectric type. An electromechanical stepper motor consists of core, coil, gear mechanism and so on. When the electromagnetic coil is energized, the magnetic force will produce, promote its core motion through a gear structure of the output shaft to rotate an angle, through the anti-rotation gear output shaft to maintain in the new position; when a coil is energized, the shaft and the rotation angle move in step motion. A magnetoelectric stepper motor mainly has three forms: permanent magnet type, reaction type and permanent magnet induction type.

3.5.4　Other special motors

Besides these special motors, there are also superconducting motors and ultrasonic motors.

The principle of electro mechanical conversion of superconducting motor is similar to that of ordinary motor, but its winding is made of superconducting material, which can greatly reduce the volume and save energy. Due to the need for superconducting refrigeration equipment, the structure is so complex that is generally used only for large generators or motors (such as the propulsion of 10 000 tons of large wheels).

A ultrasonic motor is developed in the middle of 1980s. With no magnetic field and winding, its operation principle is completely different that of from the traditional electromagnetic motor, the latter is based on the converse piezoelectric effect of piezoelectric ceramics materials, the former converts electrical energy into ultrasonic vibration of the elastomer, and will be converted to rotary friction drive or linear motion of moving object. This type of motor has the advantages of low operation speed, large power, compact structure, small volume, low noise, not interfered by the environmental magnetic field, and it can be applied to biological science, optical instruments, high precision machinery and other fields.

3.5.5   The classification history of electrical motor and electrical appliances

In a broad sense, electrical appliances refer to all implements that use electrical power. However in electrical engineering, they mainly refer to the electrical devices, equipments and components used to connect and disconnect circuits and change circuit parameters to realize the control, regulation, switching, detection and protection of circuits or electrical equipments. Motors (including transformers) are machines for the production and conversion of electrical energy. They are customarily not included in the list of electrical appliances. High voltage circuit breakers have two or three lagers in power systems, and miniature circuit breakes are equipped small ones with ordinary household switches. In the past 100 years, the general trend of electrical equipment development was capacity increase, transmission voltage increase and automation degree increase. For example, since the beginning of the 20th century, the switchgear had developed from air or oil as arc extinguishing medium, through multi oil, less oil and compressed air, to vacuum and Sulfur Hexafluoride. Its breaking capacity was about 20-30 ka from the beginning to 80-100 kA in the late of 1980s. The working voltage was increased to 1 150 kV. In 1960s, transistor relays, proximity switches, thyristor switches appeared. Then, in 1970s, there were intelligent electrical appliances with mechanical-electrical integration, as well as sulfur hexafluoride fully enclosed electrical appliances. The emergence of these electrical appliances were interdependent and mutually promoting with new electrical materials, new electrical manufacturing technology and new technology, which adapts to the continuous development of the entire electric power industry and society.

The classification of electrical appliances is listed below:

(1) The devices for connecting and disconnecting circuits, including knife switches, contactors, load switches, disconnecting switches, circuit breakers, etc.

(2) The electrical appliances for control circuit, mainly as electromagnetic starter, star delta starter, auto decompression starter, starter, frequency sensitive rheostat, control relays (starter motor is increasingly being replaced by power electronic devices).

(3) The electrical appliances for switching circuits, like transfer switches, master switches, etc.

(4) The main transformers and sensors for detecting circuit parameters of electrical appliances.

(5) The electrical appliances for protection circuits, mainly as fuses, circuit breakers, current limiting reactor and lightning arrester.

Im fact, electrical appliances includes high voltage apparatus and low voltage apparatus. AC voltage at 1 000 V and below and DC voltage at 1 500 V and below are calssified as the low-voltage electrical work; AC voltage above 1 000 V and DC voltage above 1 500 V are classified as high-voltage electrical appliances. The basic structure and function of high voltage and low voltage electrical apparatus in the power system will be introduced later.

3.6　High voltage apparatus

3.6.1　High voltage switch

High voltage switch equipments are mainly used to close and to disconnect the normal power line withdran with rated voltage of 3 000 V and above, and the transportation of switching power load; to fault equipment and fault line, to ensure the normal operation and safety of the power system; separate the two parts of the power system; and to reliably ground the equipment or lines, to ensure the safety of power lines, equipment, operation and maintenance personnel.

It includes main circuit breaker, disconnector, reclosing and contactor, fuse, load switch and other devices of the high-voltage switch cabinet. They are interdependent in structure, such as the i-solation load switch, fuse breaking combined electrical switch, etc.

The most frequently used of high-voltage switches in power systems are not circuit breakers and disconnecting switches.

A circuit breaker is a switch device that can close, load and disconnect the current in the normal circuit and under the abnormal circuit condition within the specified time. Circuit breakers are divided into high voltage circuit breakers and low voltage circuit breakers.

High voltage circuit breakers can only cut off or disconnect the load current in the high-voltage circuit. When a fault occurs, it can cut off the overload current and short-circuit current through the action of relay protection device, with a perfect arc extinguishing structure and sufficient breaking capacity, it can be divided into oil circuit breaker (bulk oil circuit breaker, oil-minimum circuit breaker) (SF6 circuit breaker, as shown in Figure 3.20), vacuum breaker and compressed air circuit breaker.

A low voltage circuit breaker, also known as automatic air switch or automatic air circuit breaker, as shown in Figure 3.21, is a manual switch function used for automatic protection against voltage loss, undervoltage, overload and short circuit. It can be used to distribute electricity, as an infrequently started asynchronous motor, protecting the power line and motor, and cut off the circuit automatically in case of serious overload or short circuit or undervoltage fault, with a function equivalent to the fuse switch and less thermal relay etc. Moreover, there is no need to change the parts after breaking the fault current. It has been widely used in the distribution system.

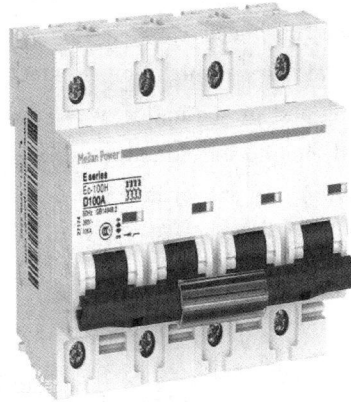

Figure 3. 20   SF6 high voltage circuit breaker   Figure 3. 21   Low voltage automatic air circuit breaker

For a disconnector, at the opening position, there is an insulation distance and obvious disconnection mark between contacts that meet the specified requirements; at the closing position, it can carry current under normal circuit conditions and current under abnormal conditions (such as short circuit) within the specified time. A disconnector (commonly known as " knife switch"), generally refers to the high-voltage disconnector, that is, the disconnector with rated voltage of 1 kV and above, is the most widely used electrical appliance in the high-voltage switchgear, as shown in Figure 3. 22. The main feature of the disconnector is that it has no arc extinguishing ability and can only open and close the circuit without load current. An isolated switch is used for all levels of voltage to change the circuit connection or isolate the circuit or equipment from the power supply. It has no breaking-capacity, so that it disconnect circuit after equipment such as circuit breaker to disconnect the line in advance. An interlocking device is usually used to prevent incorrect operation of the switch with load, and sometimes pins are required to avoid disconnecting the switch under the influence of the large faulty magnetic force.

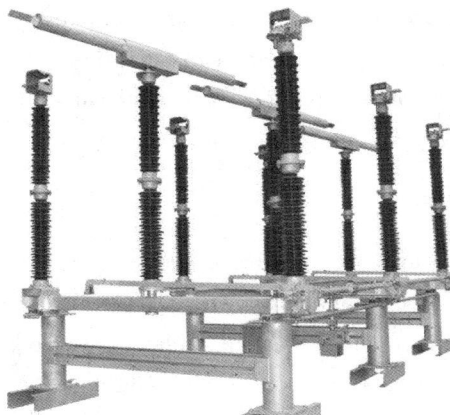

Figure 3. 22   Isolated switch

With the improvement of product sets, the single high voltage switch, and other electrical products, such as current transformer, voltage transformer, lightning arrester, capacitor, reactor, bus lines and import casing or cable terminal, are organically combined together, except the inlet and outlet. All metal cases of high-voltage electrical grounding device are completely closed, configuring with monitoring and protecting devices, consisting of a metal closed switch device with GUI control, protection and monitoring function and gas insulated switchgear (Gas insulated metal enclosed switchgear, GIS), as shown in figure 3. 23.

Figure 3. 23   Gas insulated metal enclosed switchgear

In recent years, in order to make the high voltage switchgear in a complete set, people will build the substation(including power transformer) as a whole. It will be prefabricated and debugged before delivery to the site, which can significantly reduce the workload of installation and debugging on the site, the installation and the debugging cycle are shortened greatly. It will reduce errors and deviations in the installation and commissioning work, improve the reliability of the equipment in operation. The area required for this kind of equipment is also significantly reduced, compared with the conventional equipment. Especially in the constructions of densely populated areas, such as residential areas, commercial areas, small unmanned substation are used instead of the traditional substation, not only saving the costs of construction and maintenance, but also improving the environment.

3.6.2 **Protection devices**

A lightning arrester is used to protect electrical equipments from transient over voltage hazards by limiting the duration and value of continuous currents. It is also another important equipment prevent the lightning from damaging of communication cables. Power equipments can be destructed by the operating over voltage, the internal overvoltage and lightning overvoltage.

Magnitude of lightning overvoltage and operating overvoltage may or relatively high. It is not only economically unreasonable in some way, but also technically infeasible to withstand these two kinds of overvoltage simply by increasing the insulation level of the equipment. In general, the arrester is used to limit the overvoltage to a reasonable level. With the overvoltage, the arrester will resume to the cut-off state immediately and the power system will resume its normal state immediately. The protection features of the lightning arrester is the foundation for the insulation coordination of the protection equipment. The protection reliability can be improved by improving the protection characteristics of arrester, while the insulation level of equipment can be reduced with less weight, reduce the cost.

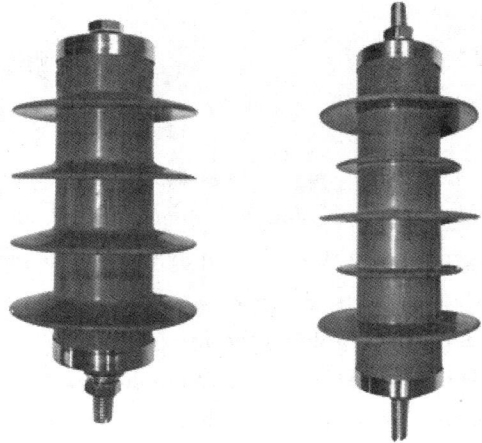

Figure 3.24   arresters

At present, the most common lightning arresters are silicon carbide valve type lightning arresters and metal oxide arresters, as shown in Figure 3.24.

3.6.3   The instrument transformer

A instrument transformer is a kind of equipment used for measuring and protecting in power system. It is divided into two categories: voltage transformer and current transformer. Instrument transformer is applied for measurement, protection and control, and to isolate high voltage devices.

The common current transformer works according to the principle of electromagnetic conversion. The structure is the same as the transformer. It is called the electromagnetic current transformer. In addition to the electromagnetic type of voltage transformer, there is also the capacitive type, as shown in figure 3.25.

The electronic transformer for measurement is widely applied.

Figure 3.25   capacitor voltage transformers

3.7 Low voltage apparatus

Low voltage apparatuses usually are the electrical equipments in the distribution and control systems which work below AC voltage 1 000 V or DC 1 500 V. It plays a role of switching, controlling, protecting, adjusting, detecting and displaying in transportation and distribution of electric energy. Low voltage apparatuses are the basic component of low voltage control loop. Low voltage electrical appliances are widely used in power plants, substations, as well as factories and mines, transportation, rural areas, buildings and other power systems. In the factory, relays, contactors, buttons and switches are often used to form the starting, stopping, reversing and braking control loops of the motor. More than 80% of the power delivered by a power plant is transmitted and distributed through various low voltage electrical appliances. According to statistics, more than 40 000 kinds of low-voltage electrical equipments are needed for each new 10 thousand kilowatts of power equipments. Therefore, with the improvement of electrification degree, the consumption of low voltage electrical appliances will increase sharply.

Figure 3.26   some low voltage transfer switches

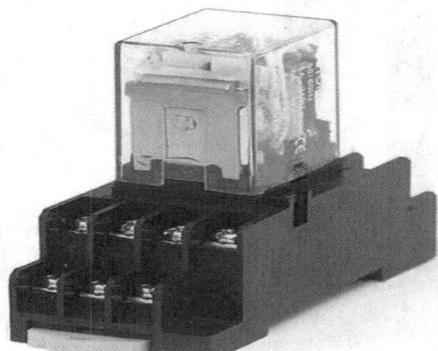

Figure 3.27   a relay

A relay is one of the most basic electrical components in modern automatic control system, as shown in figure 3.27. It is widely used in power system protection settings, such as production process automation devices, all kinds of remote power supply, remote control and communication devices. Relays can be divided into power relay (whose input is the current, voltage, frequency, power respectively) and non-electric relay (whose input is temperature, pressure or velocity), but relays have one characteristic in common, that is when the input physical quantity reaches the specified value, the electrical output circuit will be connected or disconnected automatically.

The relay contact has three basic forms:

(1) Moving contact type (normally open). When the coil is not energized, the two contacts are seperated on contrary, the two contacts will be closed once the electricity supply the coil.

(2) Moving broken type (normally closed), when the coil is not connected, the two contacts are closed, on the contrary, the two contacts will be disconnected quickly energized.

(3) Changeover type, this is the contact-group type. There are three contacts in the contact group, that is, the middle one is the moving contact and the others are the stationary contact. When the coil is not energized, the moving contact and one of the stationary contacts are disconnected, and the other is closed. After the coil is energized, the moving contact moves, so that the original opening is closed, and the original closed state is broken, so as to achieve the purpose of conversion. Such a contact group is called a changeover contact.

A contactor is a kind of electric appliance which uses the coil current to form magnetic field and make the contact close or open to control the load in industrial electricity. The contactor has large control capacity and is suitable for frequent operation and remote control. It can cut off the AC and DC main circuits quickly, and connect with large current control (some types could reach up to 800 A) circuit device frequently, so it is often used in motor as a control object. Meanwhile, it can be used as control equipments, electric heaters, machine tools and all kinds of power unit load of e-lectric contactor. It does not need to connect and cut off the circuit, with low voltage release protection. Its main control object is the motor, which can be used to control electric heating equipment, electric lighting, welding machine and capacitor group and other power load. The contactor has high operating frequency, and the highest operation frequency can reach 1 200 times per hour. The contactor has a long service life. The mechanical life is usually millions of times to ten million times operating, and the electric life is usually hundreds of thousands of times to millions of times switching.

Contactors are widely used in low voltage electrical appliances, as shown in Figure 3.28.

According to its position and function in the circuit, low-voltage electrical appliances includes: distribution device and control device, their performance requirements are also different. Adistribution equipment generally does not need frequent operation, it should have a higher breaking capacity with protective function. The control device generally does not disconnect the big currents, but its action is relatively frequent, therefore, the control device is required to have a long

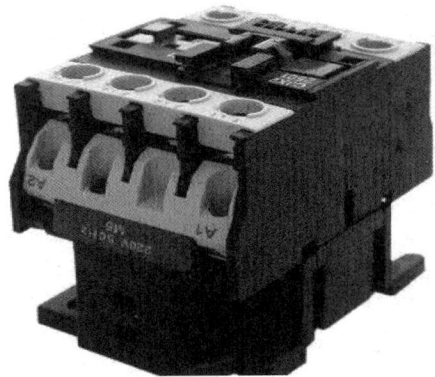

Figure 3.28　a contactor

operating life. Low voltage device can be divided into mechanical action device with contact switch and non mechanical action device according to its operation mode. Mechanical action device maybe automatic switching or non-automatic switching. The automatic switching devices operate automatically according to the changes of its own parameters or external signals, and the operation of non-automatic switching device depends on external force to complete the action.

At present, all countries in the world pay much attention to the development of low-voltage electrical apparatus, and also to the improvement of the new technology of microelectronics, new materials and new products. Especially, the great progress has been made in the intelligent, modular, integrated, multi-functional aspects of low-voltage electrical appliances.

Chapter *4*
Power Electronics

4.1　Emergence and development of power electronic technology

4.1.1　Power electronic technology

Electronic technology includes two branches: information electronic technology and power electronic technology. The electronic technology used in the field of information processing is called information electronic technology, including analog electronic technology and digital electronic technology. The electronic technology used in the field of power conversion is called power electronic technology. Power electronic technology is an emerging interdisciplinary integration of electronic technology, control technology and power technology, and its main task is to study the various power electronic devices, as well as a variety of circuits and devices compose of these power electronic devices, reasonable and efficient completion of the transformation and control of electric power. Figure 4.1 is the generic structure of a power electronic system.

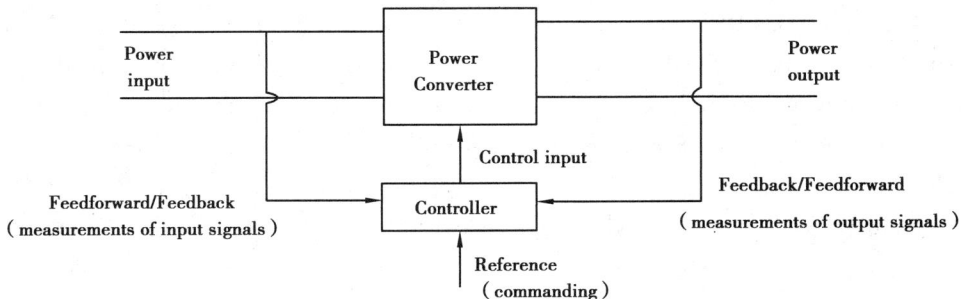

Figure 4.1　Generic structure of a power electronic system

Power electronics is the interface between electronics and power. Figure 4.2 is the interdisciplinary nature of power electronics.

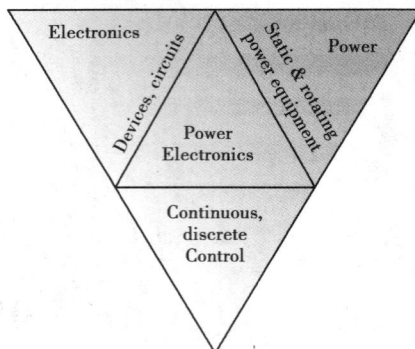

Figure 4.2   Interdisciplinary nature of power electronics

Because of the weak power of the information electronic circuit, the device can work in the switch state and the amplification state.

The power electronics devices in power electronic circuits generally are working in the switch state and should be able to flow through the large current and very low voltage drop on state withstand high voltage and very low leakage current off state; with a short time between transition state to the swithing off state and small power loss.

Power electronics technology mainly consists of 3 parts: power electronics device manufacturing technology, converter technology and control technology. Among them, the power electronics device manufacturing technology is the basis of power electronics technology, converter technology is the core of power electronic technology, and control technology is an integral part. The theoretical basis of the manufacturing technology of power electronics device is semiconductor physics. The theoretical basis of converter technology is circuit theory.

4.1.2   Development of power electronics technology

The first glass mercury arc rectifier appeared in 1902. In 1910, the iron mercury arc rectifier appeared.

The copper oxide rectifier was developed in 1920. Selenium rectifiers appeared in 1923.

In 1947, the famous Bell laboratory invented the transistor. In 1956, American J. Moore made a prototype thyristor. In 1957, American R. A. York made a practical thyristor. In the late 70s, fully controlled devices represented by the gate turn off thyristor (GTO), power bipolar transistor (BJT) and power Metal-Oxide-Semiconductor Field-Effect Transistor (Power-MOSFET) were developed rapidly.

In the late 1980s, composite devices represented by insulated gate bipolar transistors (IGBT) emerged suddenly.

4.2   Power electronics devices

Power electronics devices refer to the electronic devices that can be directly applied to the main circuit to undertake and realize the transformation or control of electric energy. Compared with the electronic information processing, it has the following characteristics: better ability of electric power, working in the switch state, the electronic circuit to control the power loss, larger power loss and installing the radiator when it works.

According to controllable level, power electronic devices are classified as below:

① The uncontrolled devices cannot control anything, so there need not driving the circuit.

② The semi-controlled device can control its turn on but not turn off by the control signals.

③ The fully controlled device can control both its turn on and turn off by the control signal, also known as the self-closing device.

According to the nature of the drive circuit signal, power electronic devices are classified as:

① A current driven type which is turned on or off by injecting or extracting current from the control side.

② A voltage driven type which can control the switch operation only by applying a certain voltage signal between the control terminal and the common terminal.

According to the waveform of the effective stimulating signals, the drive circuit is divided into two categories:

① The pulse triggering control on or off the control pulse signal by applying voltage or current on the control side, once the control pulse signal processes the turn on or blocked state and the main circuit conditions, the device can maintain its turn on or blocked state without the use of control signal.

② Level control, the device must be open and maintained on a conducting state by continuously applying a voltage or current signal, or be switched off and maintained in the blocked state.

According to the number of two kinds of carriers in the device, namely the free electrons and holes, are involved in the conduction channel, and they are divided into three types:

① Unipolar device, a device with a single carrier participating in the conduction.

② Bipolar device, a device consisting of both free electrons and holes with two kinds of carriers participating in the conduction.

③ Hybrid device, a device consisting of composite unipolar devices and bipolar device.

4.2.1   Uncontrollable device

The structure and principle of Power Diode are simple and reliable. It has been applied since the early 1950s. The basic structure and working principle are the same as those of diodes in microelectronic circuits. It consists of a large PN junction, two lead ends and a package. There are two kinds of package: bolt type and flat type. The common power diode, also called rectifier diode, mainly owns the unidirectional conduction ability of PN junction, which has large capacity and long reverse recovery time. It is mostly used in rectifier circuit with low switching frequency (less than 1 kHz).

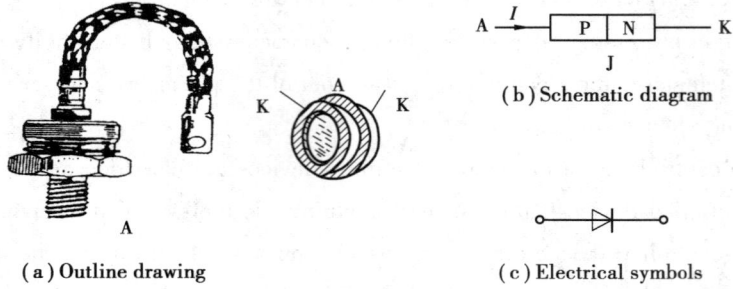

(b) Schematic diagram

(a) Outline drawing

(c) Electrical symbols

Figure 4.3   Power diode

Figure 4.4   power diode

4.2.2   Semi controlled device

The thyristor is referred to as the crystal thyratron, also known as the silicon controlled rectifier (SCR). As a kind of typical small current control current elements controllable in the working process and able to work under high voltage and big current conditions, it is widely used as controllable rectifier, AC voltage, no-contact electronic switch, inverter in electronic circuits. In 1957, General Electric Company developed the world's first thyristor product, and commercialized it in 1958. The thyristor as shown in Figure 4.5.

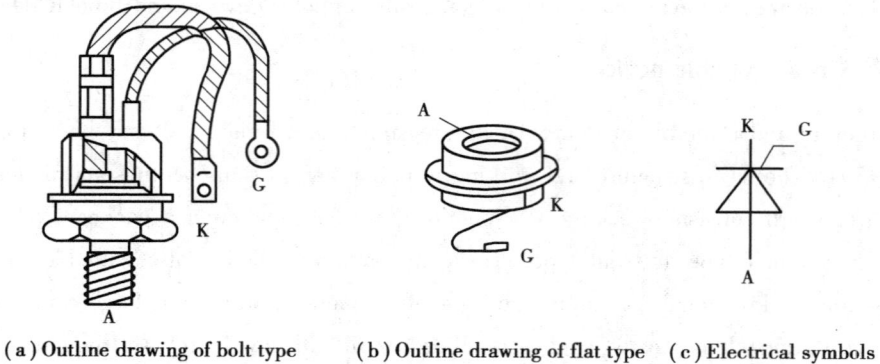

(a) Outline drawing of bolt type   (b) Outline drawing of flat type   (c) Electrical symbols

Figure 4.5   thyristor

4.2.3　**Fully controlled device**

Fully controlled devices include power transistors (Figure 4.6), power field effect transistors, insulated gate bipolar transistors and so on.

① Power Giant Transistor(GTR)

GTR is actually the bipolar junction transistor that can handle high voltage and large current. So GTR is also called power BJT, or just BJT.

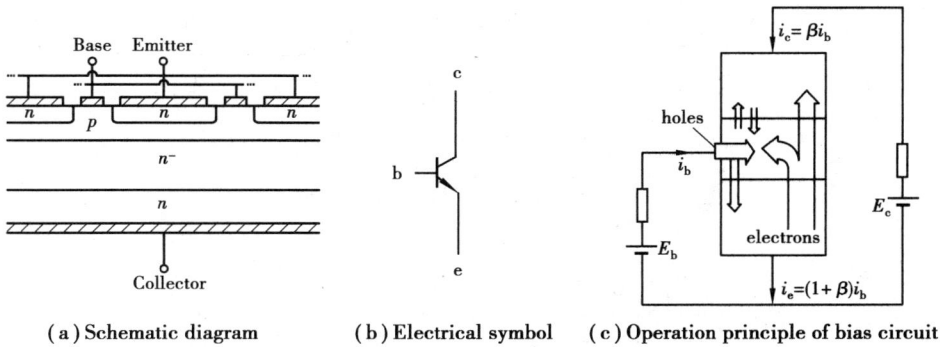

(a) Schematic diagram　　(b) Electrical symbol　(c) Operation principle of bias circuit

Figure 4.6　power transistors

② Power Metal Oxide Semiconductor field effect transistor (Power MOSFET)

MOSPET are divided into junction type and insulated gate type. Usually they are referred to as insulated gate field effect transistors. Power field effect transistors can be divided into P channel and N channel according to the conductive carriers. According to the relationship between the gate source voltage and the conduction channel, the power field effect transistor can be divided into depletion-mode type and enhancement-mode type. Power field effect transistors are generally N channel Enhancement mode.

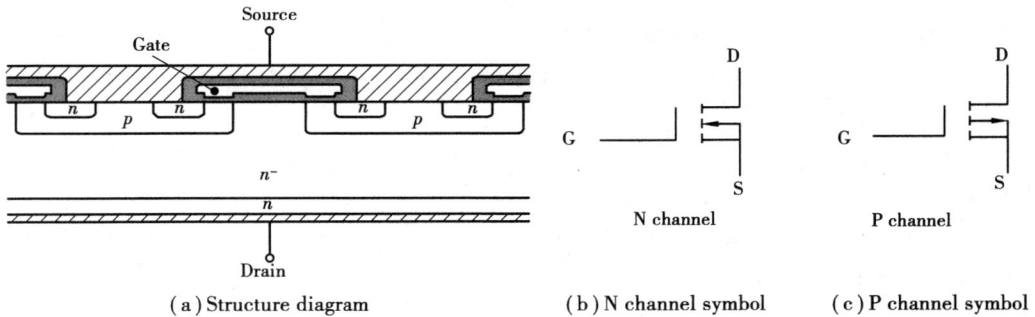

(a) Structure diagram　　　　(b) N channel symbol　　(c) P channel symbol

Figure 4.7　P-MOSFET Structure and electrical symbols of a unit

③ Insulated gate bipolar transistor(Insulate-Gate Bipolar Transistor—IGBT)

IGBT integrate power transistors(Giant Transistor—GTR) are similar to the field effect transistor (Power MOSFET). It has many advantages, such as high input impedance, fast switching speed, simple drive circuits, low on state voltage, high voltage and high current, and so on. It has been widely used in frequency converter and other speed regulating circuits. IGBT is also a three-side device with gate G, collector C and emitter E.

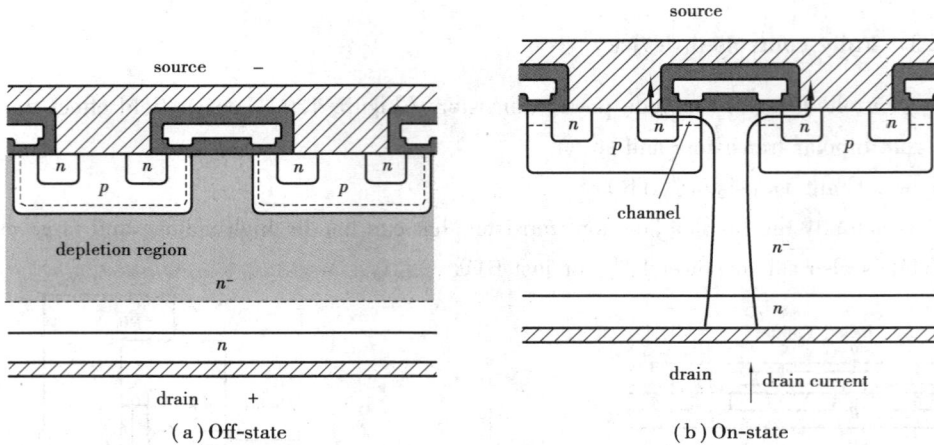

Figure 4.8　Physics of MOSFET operation

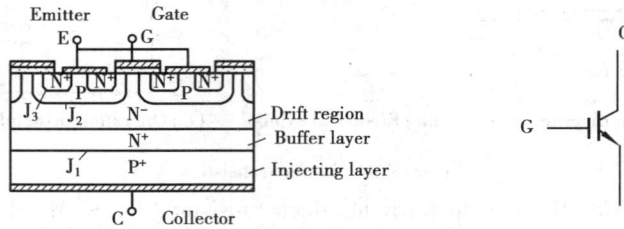

Figure 4.9　IGBT and its electrical symbols

4.2.4　Other new power electronic devices

① MOS Controlled thyristor(MCT)

Essentially a GTO with integrated MOS-driven gates controlling both turn-on and turn-off that potentially will significantly simplify the design of circuits using GTO. The difficulty is how to design a MCT that can be turned on and turned off equally well.

② Static induction transistor (SIT): Namely, power junction field effect transistor-power (JFET).

③ Static induction thyristor(SITH): Namely, field controlled thyristor-FCT.

④ Integrated gate commutated thyristor(IGCT).

Actually, the close packaging of GTO and the gate drive circuit with multiple MOSFETs in parallel provide the gate currents.

4.2.5　Power module and power integrated circuit

The device with logic operation, control, protection, sensing, detection and self diagnosis information of electronic circuits integrated on the same chip is called power integrated circuit (PIC).

There are many types of power electronics circuits that encapsulate the way of integration.

Smart power integrated circuit (SPIC) usually refers to the integration of longitudinal power devices with control and protection circuits. It is commonly used in voltage regulator, automobile power switches, switching power supply, motor drives, household appliances and other products.

High voltage integrated circuit (HVIC) generally refers to the monolithic integration of high voltage devices with logic or analog control circuits. It is usually used in small motor drives and telephone switches, user circuits and other places requiring higher voltage.

Intelligent power module (IPM) refers to the monolithic integrated of Insulated Gate Bipolar Transistor (IGBT) and its auxiliary device protection and drive circuit.

Figure 4. 10   power electronic module

Integrated power electronic module and power integrated circuit are specific power electronic integration technology. Power electronic integration technology can bring a lot of benefits, such as reducing device volume, improving reliability, convenience for users, reducing manufacturing, installation and maintenance costs. It has a wide application prospect.

4.3   Power electronic converter technology

Converter technology mainly includes: the use of power electronic devices to constitute a variety of power conversion circuits, the control of the circuit, and the use of these technologies to constitute a more complex power electronic devices and systems. Commonly, "converter" refers to: alternating current DC (AC-DC), DC alternating current (DC-AC), DC variable DC (DC-DC) and alternating current (AC-AC).

4.3.1   AC-DC converter

Rectifier circuit is the earliest power electronic circuit to change the AC into DC power supply. The rectifier circuit is usually composed of main circuit and control circuit, as shown in Figure 4. 11. The main circuit includes transformer, rectifier and filter.

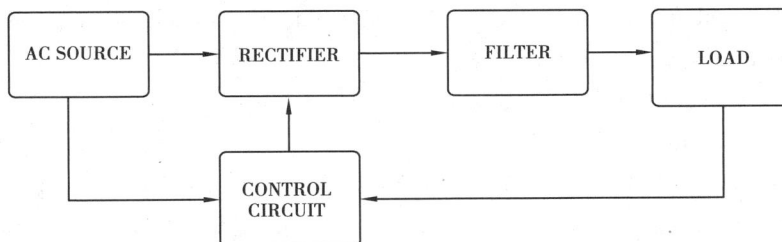

Figure 4. 11   modules of AC-OC converter

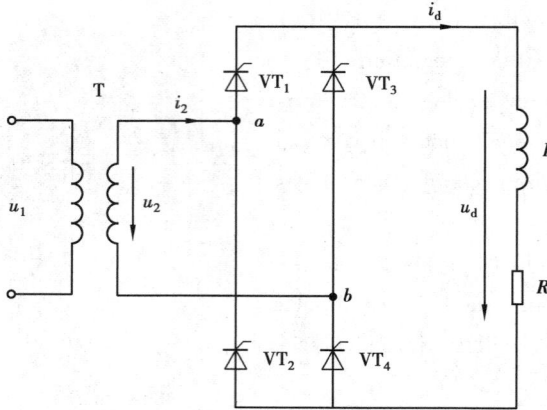

Figure 4.12   single phase bridge controlled rectifier circuit

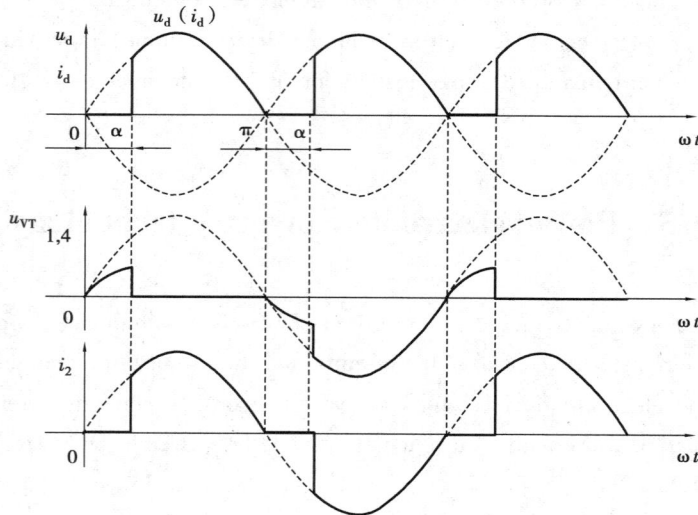

Figure 4.13   circuit waveform of resistive load

$$U_d = \frac{1}{\pi} \int_{\alpha}^{\pi+\alpha} \sqrt{2} U_2 \sin \omega t d(\omega t) = \frac{2\sqrt{2}}{\pi} U_2 \cos \alpha = 0.9 U_2 \cos \alpha \qquad (4.1)$$

Rectifier circuit is mainly used in the following fields for demand of OC power suply:

① Electrochemical treatment. For example: electroplating, metal refining and chemical gas (hydrogen, oxygen, chlorine, etc.) production.

② Adjustable speed DC drive system and AC drive system.

③ HVDC transmission system.

④ Universal AC / DC power supply, including an uninterruptible power supply system.

⑤ New energy power generation technology. For example: solar photovoltaic power generation, wind power, fuel cells and other energy conversion circuit.

4.3.2 DC-AC transforms

DC-AC (inverter) is the process of converting DC battery into AC 220 V (50 Hz).

(a) Motorcycle rectifier

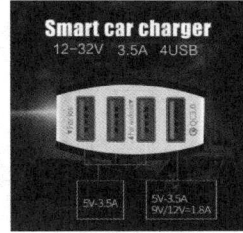

(b) Smart car charger

Figure 4.14   application of rectifier circuit

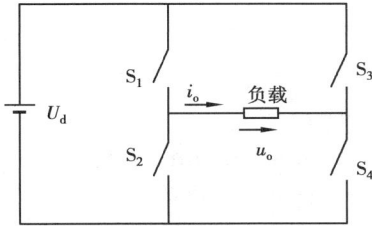

(a)

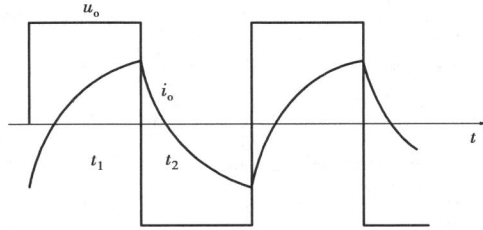

(b)

Figure 4.15   inverter circuit(a) and its waveform example(b)

A complete inverter circuit consists of four parts: inverter main circuit, drive and control circuit, input / output circuit and auxiliary circuit.

(1) Inverter main circuit

The inverter main circuit, composed of power electronic switches, is the main part of energy conversion.

(2) Drive and control circuit

The drive and control circuit mainly generates a series of control signals according to the control requirements, realizes the control of the main circuit, and completes the voltage regulation, frequency modulation and other functions of the inverter circuit.

(3) Input / output circuit

The input circuit is mainly DC, and the input circuit must be set in order to ensure that the DC power supply is a constant voltage source or constant current source original energy storage. Capacitor elements are used to stabilize voltage, and inductors are used to stabilize current. The output circuit is mainly filter circuit.

(4) Auxiliary circuit

The auxiliary circuit includes auxiliary power supply and protection circuit. The auxiliary power supply provides DC operating voltage for each part of the circuit. The protection circuit is mainly executed with the failure of the circuit under normal circumstances.

4.3.3   AC-AC converter

An alternating current converting circuit is to convert one form of alternating current into another one. The related voltage (current), frequency and phase can all be changed.

AC converter circuit can be divided into two categories: one is to change the level of the voltage

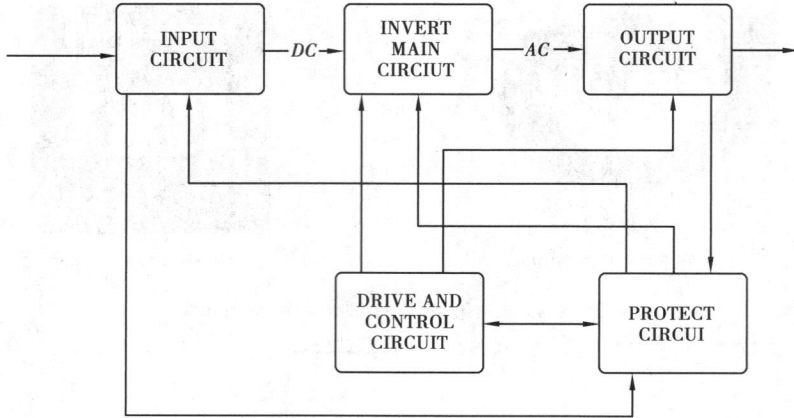

Figure 4.16　Basic structure of inverter circuit

or the implementation of the circuit on-off control, without changing the frequency of the circuit, known as AC power control circuit. The other is to turn power frequency in to another fixed or variable frequency AC power, known as the frequency conversion circuit. According to the different control methods, the AC power control circuit is divided into AC voltage regulator circuit, AC power regulator circuit and AC contactless switch.

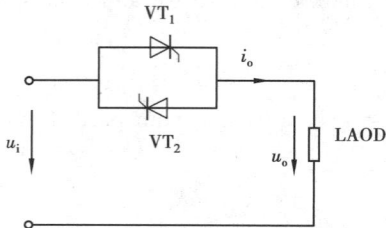

Figure 4.17　AC voltage regulating circuit

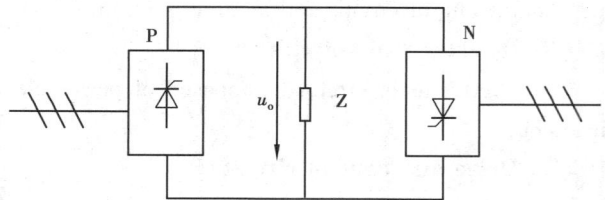

Figure 4.18　AC-AC conversion circuit

(a) Frequency converter

(b) Voltage regulator

Fig 4.19　Application of AC converter circuit

4.3.4　DC-DC converter

DC chopper is a kind of power electronic circuit that converts the fixed DC voltage to the adjustable DC voltage. It is widely used in DC switching power supply and DC motor drive system.

Chopper circuit has 6 basic circuits: Buck chopper circuit, boost chopper circuit, Boost—Buck chopper circuit, Cuk chopper circuit, Sepic chopper circuit and Zeta chopper circuit.

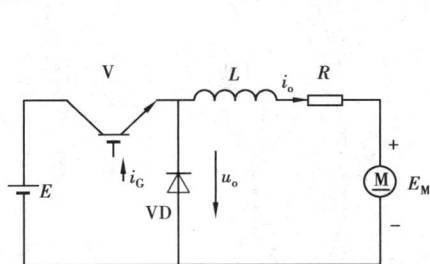

Figure 4.20   buck chopper circuit

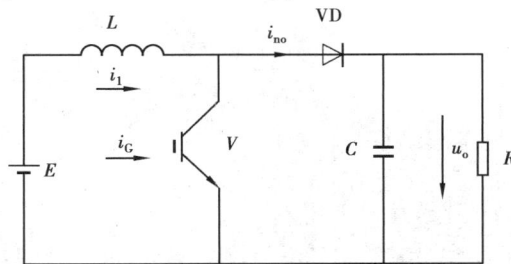

Figure 4.21   boost chopper circuit

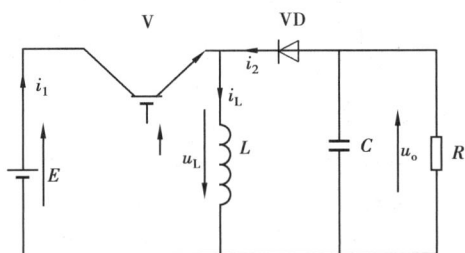

Figure 4.22   buck chopper circuit

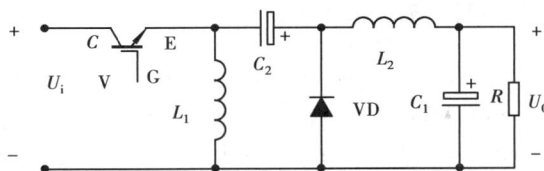

Figure 4.23   Sepic chopper circuit

4.4   Electric drive technology

Electric drive technology refers to the use of electric drive motor to convert electric energy into mechanical energy to drive all kinds of production machinery, vehicles and life technologies needed for sports projects. Through the rational use of electrical equipment and system technology, the production process control of mechanical equipment electrification and automation is realized.

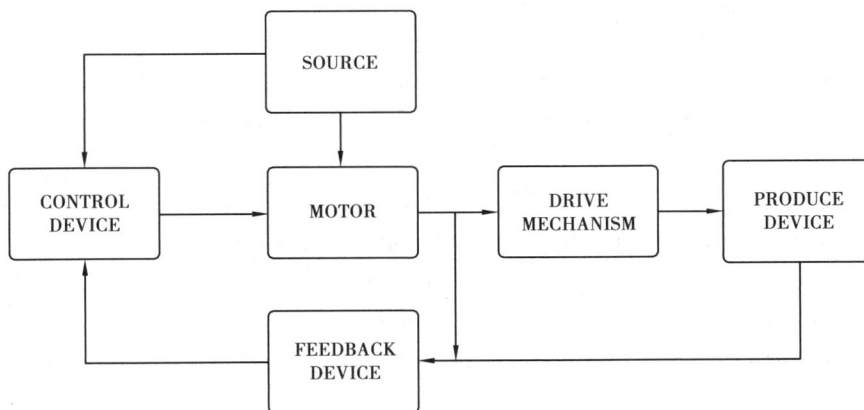

Figure 4.24   electric drive system

4.4.1 **Application of electric drive**

(a) High-speed railway

(b) Elevator

Figure 4.25   Application of electric drive

4.4.2 **DC motor drive**

DC motor is a kind of motor driven by direct current, which is widely used in small electrical apparatus. The working principle of DC motor is as follows: when the coil is energized, the magnetic field is generated around the rotor, the left side of the rotor is pushed away from the left magnet, and attracted to the right side, resulting in rotation. The rotor continues to rotate by inertia. When the rotor runs in a horizontal position, the current converter reverses the current direction of the coil and reverses the magnetic field produced by the coil, making the process repeated.

The expression of rotational speed omega of DC motor is:

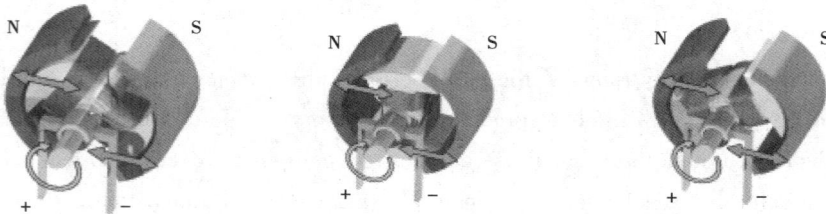

Figure 4.26   Working process of DC motor

The armature voltage is U_a; I_a is the motor armature current, R_a is the motor circuit resistance; C_e is a motor with motor potential constant; φ is the motor exciting flux.

$$\omega = \frac{U_a - R_a I_a}{C_e \varphi}$$

Through the formula, there are three DC motor speed control methods:

① Voltage regulation and speed regulation, keeping R_a and phi unchanged, and adjusting n by adjusting U_a, is a large range stepless speed regulation mode.

② Weak magnetic rising speed, keeping R_a and U_a unchanged and increasing N by decreasing the value of V, is a small range stepless speed regulation mode.

③ Variable resistance speed control, maintaining U_a and phi unchanged, by adjusting the R_a to adjust the N, is a large range of speed regulation.

4.4.3 AC motor drive

Alternating current motor is a machine that transforms the electric energy of alternating current into mechanical energy. AC motor is mainly made up of an electromagnet winding or distributed stator winding and a rotating armature or rotor used to generate a magnetic field, which is made by the forced rotation of the energized coil in the magnetic field.

AC motors are divided into two categories: asynchronous motor and synchronous motor. According to the basic principle of asynchronous motor, the electromagnetic power Pm introduced from the stator can be divided into two parts:

A part of the effective power to drag the load:

$$P_1 = (1 - s)P_m \tag{4.2}$$

The other part is the slip power:

$$Ps = sP_m \tag{4.3}$$

4.4.4 Development direction of electric drive technology

The general development trend of electric drive automation technology is that AC frequency conversion speed control is gradually replacing DC speed control and contactless control rather than contact logic control; full digital control and digital analog compound control coexist. The development of electrical automation technology is driven by technological needs of users and related subjects, directly relative to the improvement of power transmission performance, price, size, energy consumption and saving design, debugging etc. Its main development direction is:

① Realization of high level control.

② Converter for developing clean electric energy.

③ Systematization.

④ CAD technology.

⑤ Reduction of the size of the device.

Chapter 5

Power System and Automation

5.1 Electrical power industry

The electrical power industry is an industrial sector that transfers coal, oil, natural gas, nuclear fuel, power, ocean energy, wind energy, solar energy, biomass energy and other energy into electricity by power generation facilities, and then provides customers electricity through the transmission, substation and distribution system.

Electricity production and consumption process simultaneously, and they can't be interrupted. The electricity can't be stored on a large scale due to the limit of current technology.

Therefore, power operation needs unified scheduling and planning which is very important for power enterprises.

Figure 5.1 Dispatching Center

The power industry which provides the fundamental power for industry and other sectors of the national economy is the leading sector of the development of the national economy.

The power industry mainly includes five production sectors:

① Power generation, including thermal power generation, hydro power generation, nuclear power and other new energy generation technologies.

② Transmission: including AC transmission and DC transmission.

③ Substation: transformation of voltage, including step-up transformer and step-down transformer.

④ Distribution: collection and distribution of electrical energy.

⑤ Power consumption: including the installation and use of power equipment and power load control.

The power industry also includes the following sectors: planning, survey and design, construction, operation, maintenance, safety supervision, scientific research and development, equipment manufacturing, education and training, regulations and standards, power marketing (market),etc.

5.1.1 Development of world power industry

In 1875, the world's first thermal power plant was built to supply power for the nearby lighting in the Northern Railway Station in Paris, France.

In 1879, the American power plant in San Francisco to generate electricity, became the world's first power plant to sell electricity.

In the 1880s, the world's first batch of hydropower stations were built in the United Kingdom and the United states.

In 1913, the annual electricity output of the whole world reached 50 billion kW. h. From then on, the electric power industry had become an independent industrial sector, and it began to enter the field of human production and life.

In the 20th Century, 30s and 40s, the United States had become an advanced country in the power industry, with 31-generation sets of 200 thousand kW and 9 medium-sized thermal power plants of 300 thousand kW. During the same period, the hydropower generating units reached 50-100 thousand kW.

In 1934, Grand Coulee Hydroelectric Power Station was built in the USA, which had been the world's largest hydropower station before the middle of the 80 s. It began generating electricity in 1941 and got an annual installed capacity of 6.49 million kW in the 80 s.

In 1950, the world's electricity output increased to 958.9 billion kWh, 19 times as much as in 1913. In the 50s, 60s and 70s, the average annual electricity output growth rates were 9.4%, 8% and 5.3% respectively.

In 1950-1980, electricity output increased by 7.9 times, with an average annual growth rate of 7.6%, and it doubles every 10 years.

In 1986, in the world total power generation, hydropower accounted for 20.3%, thermal power accounted for 63.7%, and nuclear power accounted for 15.6%.

In the world:

the US hydropower accounted for 11.4%, thermal power accounted for 72.1%, nuclear power accounted for 16%;

the former Soviet Onion hydropower accounted for 13.5%, thermal power accounted for 76.4%, nuclear power accounted for 10.1%;

Japanese hydropower accounted for 12.9%, thermal power accounted for 61.8%, nuclear power accounted for 25.1%;

Chinese hydropower accounted for 21%, thermal power accounted for 79%.

In the 1970s, the power industry entered into the development stage of large units, large power plants, EHV and UHV transmission lines, forming a new era characterized by the United system.

In 1973, cross turbine generator units of 1.3 million kW produced by Swiss BBC company were put into operation in the United States Kenbolan power plant.

In 1981, the Soviet Union manufactured the world's largest capacity of single shaft turbo generator set of 1.2 million kW and put it into operation.

By 1977, the United States had owned 120 large thermal power plants with a capacity of more than one million kW.

In 1985, the Soviet Union had 59 thermal power stations of millions of kW.

In 1983, Japan had 32 thermal power stations above million kW, in which Kagoshima power plant became the world's largest oil-fired power plant with a total capacity of 4.4 million kW.

During this period, the largest hydropower station designed in the world was Itaipu hydropower station in Brazil and Paraguay, with a design capacity of 12.6 million kW, installed capacity of 4.9 million kW and units of 700 thousand kW. It had the same capacity as the world's largest water turbine group of 700 thousand kW in the world's largest hydropower station -Grand Coulee hydropower station.

In August 1989, ±500 kV Chinese Gezhouba Dam-Shanghai transmission line with a length of 1080km was put into operation.

In 2014, two ± 800 kV HVDC transmission projects in Zhengzhounan-Hami and Xiluoduzuo-an-Zhejiang Jinhua were put into operation.

Not only used for long-distance and large capacity transmission of electric energy, but UHV AC/DC transmission also plays an important role in the joint power system of large industrial countries or a unified power system in China.

5.1.2　Power industry in China

On July 26, 1882, the Shanghai Electric Company founded by British businessman R. W. Little, began to generate electricity. The plant installed a 16 horsepower steam generator and 15 Arc lamp to start the electricity application in China.

The application of electrical energy in China is almost synchronized with Europe and America and slightly earlier than Japan.

After that, foreign capital had successively opened some electric power enterprises in Tianjin, Wuhan, Guangzhou and other places. China's capital began to invest in the power industry in 1905 with a slow growth rate.

Before the establishment of PRC(People's Republic of China), the highest China's power generation capacity was 5. 96 billion kW / h in 1941, and China's power generation equipment capacity arrived at 1. 85 million kW, with a power generation of 4 . 31billion kW / h in 1949.

After the establishment of PRC, thermal power units with a capacity of 600 thousand kW had started running. China's large investment in the power industry led the great development of the power industry, and +100MW capacity units of the thermal power had reached 30 . 94 million kW.

By the end of 2007, the installed capacity of China's electric power was 700 million kW. In China, the transmission line with a length of 220 kV and above was 333. 8 thousand km, increased 17. 45% ; the capacitance of 220 kV and above was 1. 16 billion kVA, increased 19. 59%. The scale of electric power construction remains at a historically high level.

At present, achievements of China's power industry in the world ranked No. 1 are as follows:

① The largest power grid.

② The installed capacity.

③ The world's highest transmission voltage level.

④ The world's largest power output capacity.

⑤ The world's largest thermal power units of million kW.

⑥ The longest UHV transmission line in the world.

⑦ The earliest operation of ultra-supercritical air cooling units of million kW class.

⑧ The largest hydropower installation.

⑨ The largest hydropower station.

⑩ The highest double curvature arch dam.

5.2   Composition of power system

The power system is a whole of power generation, transmission, substation, distribution and power supply, combined by power lines, substations and power consumers at all levels of voltage power lines.

The power system equipment includes primary equipment and secondary equipment.

(1) **Primary equipment (Main Components)**

Primary equipment refers to the electrical equipment directly used in the production, transportation and distribution of electrical energy production process.

It includes generator, transformer, circuit breaker, earthing switch, disconnector, lighting arrester, bus, overhead line, power cable, reactor, etc.

(2) **Secondary equipment (Support Components)**

Secondary equipment refers to the low voltage electrical equipment needed for providing operating conditions or production command signals for operation and maintenance personnel and conducting monitoring, control, adjustment and protection of the work of the primary equipment.

Such as fuse, control switch, relay, watt-hour meter, control cable, etc.

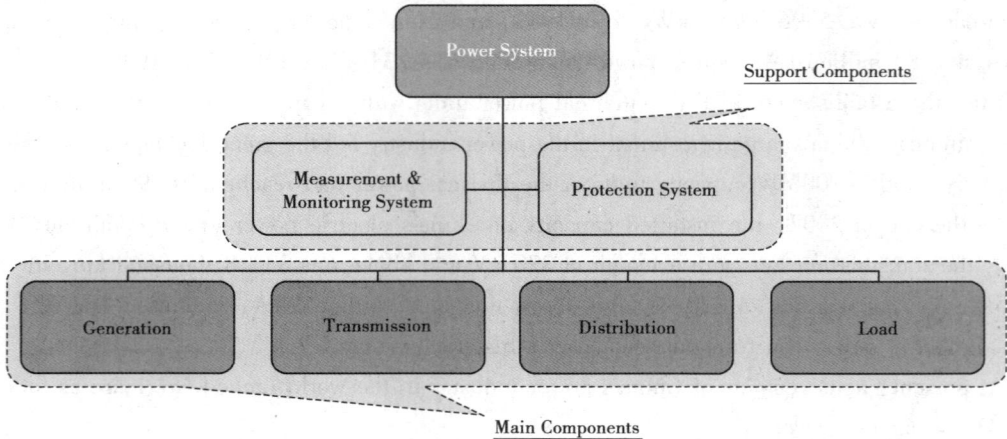

Figure 5.2　Composition of power system

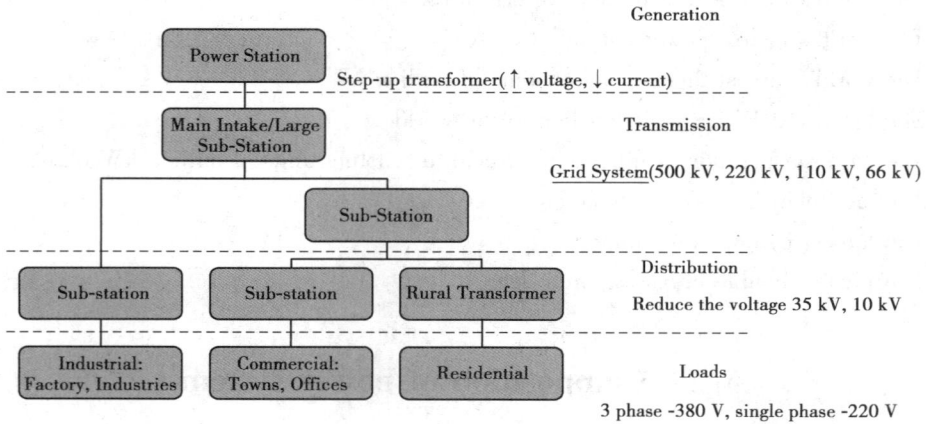

Figure 5.3　Power system diagram

　　Power lines with all levels of voltage and their related substations in the power systems are called the power grid.

　　Composed by the transmission lines of different voltage levels and transformers, the power network can be divided into three types: local power network, regional power network and EHV long distance transmission network.

　　① Local power network: power network with voltage below 110 kV (such as 35 kV, 10 kV, etc.).

　　② Regional power network: power network with voltage over 110 kV (such as 220 kV).

　　③ EHV long distance transmission network: power network with a voltage of 330-500 kV and above.

Layout of power industry in China

　　On April 12, 2002, the State Council issued the "reform plan of electric power system" (referred to as "No. 5 article"), which was regarded as the symbol of the beginning of the reform of the power system. The three core parts of the new plan are: "implementing the separation of the fac-

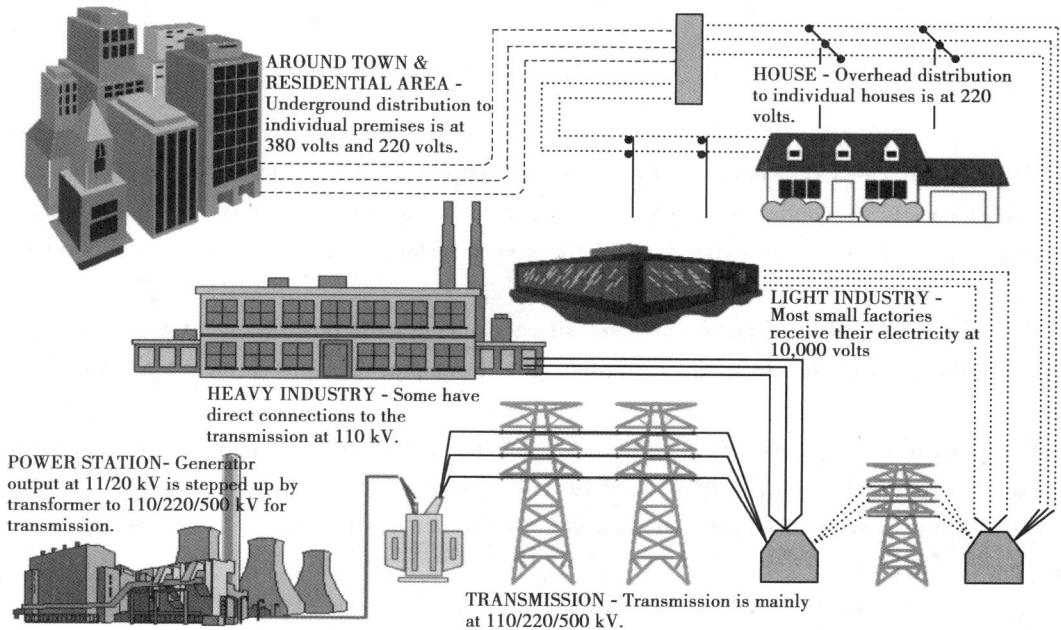

AROUND TOWN &
RESIDENTIAL AREA -
Underground distribution to
individual premises is at
380 volts and 220 volts.

HOUSE - Overhead distribution
to individual houses is at 220
volts.

LIGHT INDUSTRY -
Most small factories
receive their electricity at
10,000 volts

HEAVY INDUSTRY - Some have
direct connections to the
transmission at 110 kV.

POWER STATION- Generator
output at 11/20 kV is stepped up by
transformer to 110/220/500 kV for
transmission.

TRANSMISSION - Transmission is mainly
at 110/220/500 kV.

Figure 5.4　Power system

tory network and bidding for the Internet; restructuring the power generation and power grid enterprises; splitting the state power company vertically and horizontally. Initial establishment of competitive and open regional electricity market". Therefore, the former State Power Grid set up five power groups, two Power Grid Corps and four power assistant industry groups according to the principle of separation of factory and network. On December 29, 2002, twelve enterprises and units involving into power reform, comprised by the State Electricity Regulatory Commission and other relevant enterprises, were established. The twelve reformed units which listed simultaneously, included the State Electricity Regulatory Commission, the state and the southern Power Grid Corp, five power generation groups and four assistant industry groups.

The five major power generation groups are China Guodian, Huaneng Group, China Datang Corporation, China Huadian Corporation and China Power Investment Corp.

Four assistant industry groups are formed by two design units which are the institute of hydropower planning and design & Institute of power planning and design, and two construction units which are Gezhouba Dam group and construction corp of water conservancy and hydropower.

The national Power Grid Corp has 5 regional Power Grid Corp, including Northern China (including Shandong), Northeast (including East Inner Mongolia), East China (including Fujian), central China (including Sichuan, Chongqing) and northwest china. The Southern Power Grid Corp is composed of five power grid of Guangxi, Guizhou, Yunnan, Hainan and Guangdong.

5.3 Power plants

Power plants, also known as power stations, are the plants that convert all kinds of primary energy into electricity (secondary energy). At the end of the nineteenth century, with the growth of power demand, people began to put forward the idea of establishing the electric power production center.

With the development of motor manufacturing technology, the electric energy application scope expanded, the demand for electricity increased, the development of power plant was also accelerated.

The present power plants have multiple ways to generate electricity: thermal power plants which use coal, oil or natural gas to drive steam turbines and generate electricity, hydropower stations which use hydropower, and power stations which use new energy technology, such as wind, solar, tidal, geothermal, etc.

Types of power plants: Classification by the "mechanical means" used to drive the generator…, they are:

Thermal (water steam by burning Coal, Oil, NG) plant

Hydroelectric (falling water)

Nuclear (water steam by Uranium or Plutonium fission)

Solar

Wind

Biomass

Geothermal

…

5.3.1 Thermal power generation technology

Thermal power plants are the ones that use coal, oil and natural gas as fuel to generate electric energy.

The basic process of a thermal power plant is as follows: combustion of fuel in the boiler converts water into high temperature and high-pressure steam and converts the chemical energy of fuel into heat energy. High-pressure steam makes the turbine rotate and converts heat energy into mechanical energy, and then the turbine rotates the generator to convert mechanical energy into electrical energy.

The main equipment of thermal power plants are boilers, steam turbines and generators, and the main systems of it include:

① Steam water system. It is composed of boiler, steam turbine, condenser, water pump, heater and pipeline.

② Fuel and combustion system. It is composed of a coal handling system, pulverizing system, smoke air system and ash and dust removal system.

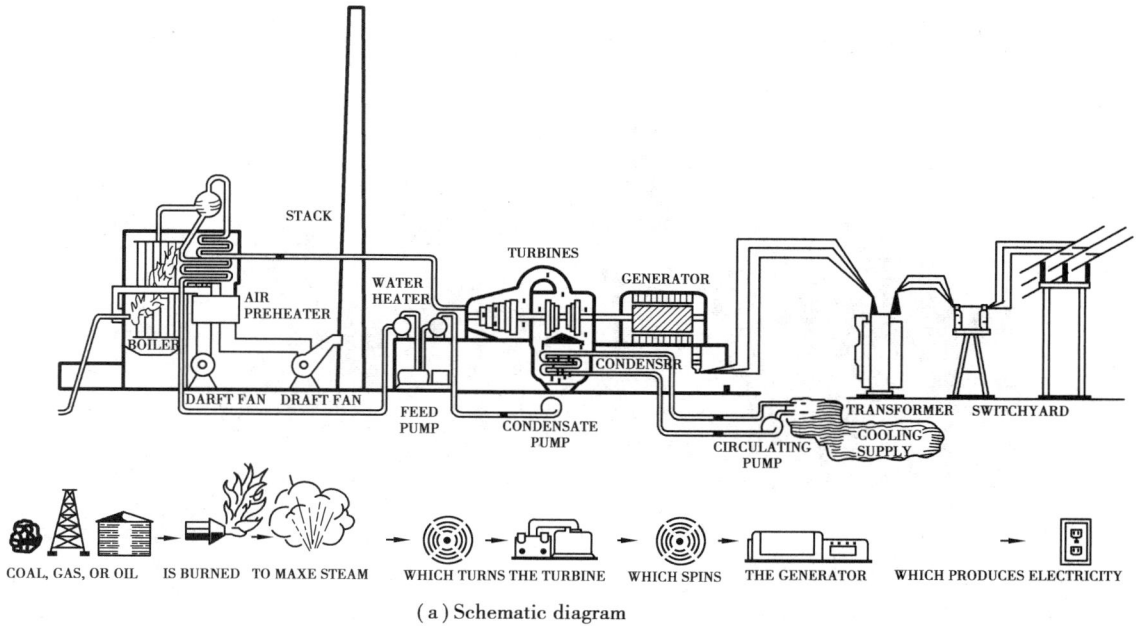

(a) Schematic diagram

COAL, GAS, OR OIL   IS BURNED   TO MAXE STEAM   WHICH TURNS THE TURBINE   WHICH SPINS   THE GENERATOR   WHICH PRODUCES ELECTRICITY

(b) appearance view

Figure 5.5   Thermal power plant

③ Other assistant thermodynamic systems.

④ Electrical system.

5.3.2   Hydroelectric power generation technology

Water is the source of human life, including natural water (rivers, lakes, atmospheric water, seawater, groundwater, etc.), distilled water and artificial water (water combined by hydrogen and

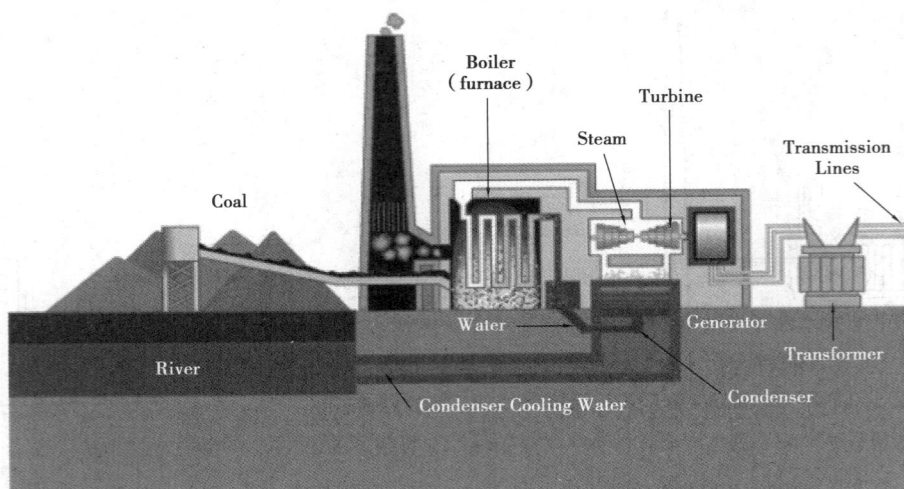

Figure 5.6 Coal-fired power plant

oxygen atoms through chemical reactions). In a broad sense, water energy resources include river water energy, tidal water energy, wave energy and ocean current energy resources; in a narrow sense, water power resources means river water resources. It is conventional and primary energy. At present, the most mature and available hydropower is river energy.

Water power, also called the hydroelectric power, is renewable energy. It is one way to generate electricity by converting potential energy and kinetic energy into electricity. The plants which generate electricity by hydropower are called hydroelectric plants, or hydropower plants, or hydropower stations. Water power is mainly used in hydroelectric power generation, with its advantages of low cost, continuous regeneration, no pollution, and disadvantage of limits by natural conditions such as hydrology, climate and geomorphology.

There are different classification methods for hydropower stations.

According to the nature of water utilization of hydropower plants, they can be divided into three types:

① Conventional hydropower plants: to use natural rivers, lakes and other water sources to generate electricity.

② Pump & storage power plants: to pump the water of low reservoir to the high reservoir for storage in the use of excess load in the period of down power load, draw off water to generate electricity in the period of peak power load , and lead the tailwater flowing into the low reservoir to meet the need of the peak power load, etc.

③ Tidal power plants: to use tidal power formed by ocean tide fluctuations to generate electricity.

According to the utilization and regulation ability of natural flow, the hydropower plants can be divided into two types:

① Run-of-river-type hydropower plant: hydropower plant with no or little reservoir capacity and with no or little regulating capacity for natural water;

② Water-storage-type hydropower plant: hydropower plant with a reservoir with certain storage capacity and with different regulating ability to natural water flow.

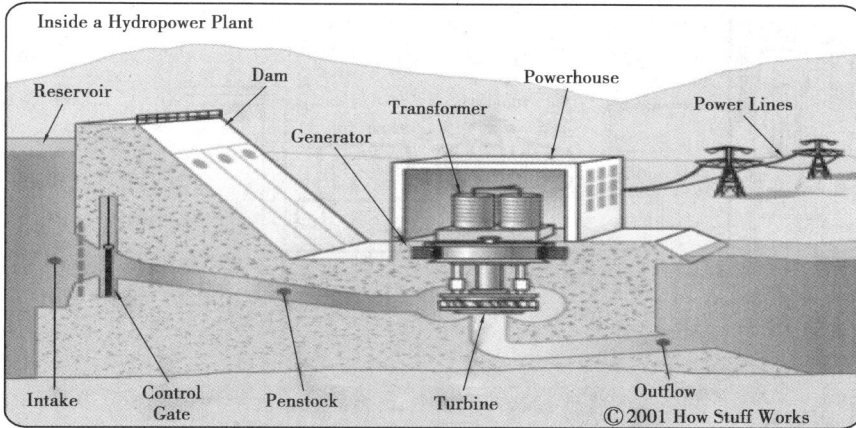

(a) Schematic diagram

(b) appearance oiew

Figure 5.7   Hydro power plant

5.3.3   Nuclear power generation technology

Nuclear energy, also called atomic energy, comes from nuclear fission or fusion. A typical nuclear fission formula is listed as follows:

$$^{235}_{92}U + ^{1}_{0}n \rightarrow ^{142}_{56}Ba + ^{91}_{36}Kr + 3^{1}_{0}n \tag{5.1}$$

Nuclear fission can release heat, which can be used to generate electricity. Therefore, it is an important method to get low carbon power.

Nuclear power generation uses the heat generated by a nuclear fission chain reaction of uranium fuel to heat water into high temperature and high-pressure steam turbine, which converts heat into mechanical energy, and the generator converts mechanical energy into electrical energy.

5.3.4   Solar power generation technology

New energy gets its name in the comparison of the conventional energy generation technology.

STEAM GENERATOR   STEAM GENERATOR
PRESSURIZER
WATER HEATER
TURBINES
GENERATOR
COOLANT PUMP   COOLANT PUMP
REACTOR
CONDENSER
FEED PUMP
CONDENSATE PUMP
CIRCULATING PUMP
COOLING SUPPLY
TRANSFORMER   SWITCHYARD

ATOMIC REACTION MAKES HEAT TO MAKE STEAM   WHICH TURNS THE TURBINE   WHICH SPINS   THE GENERATOR   WHICH PRODUCES ELECTRICITY

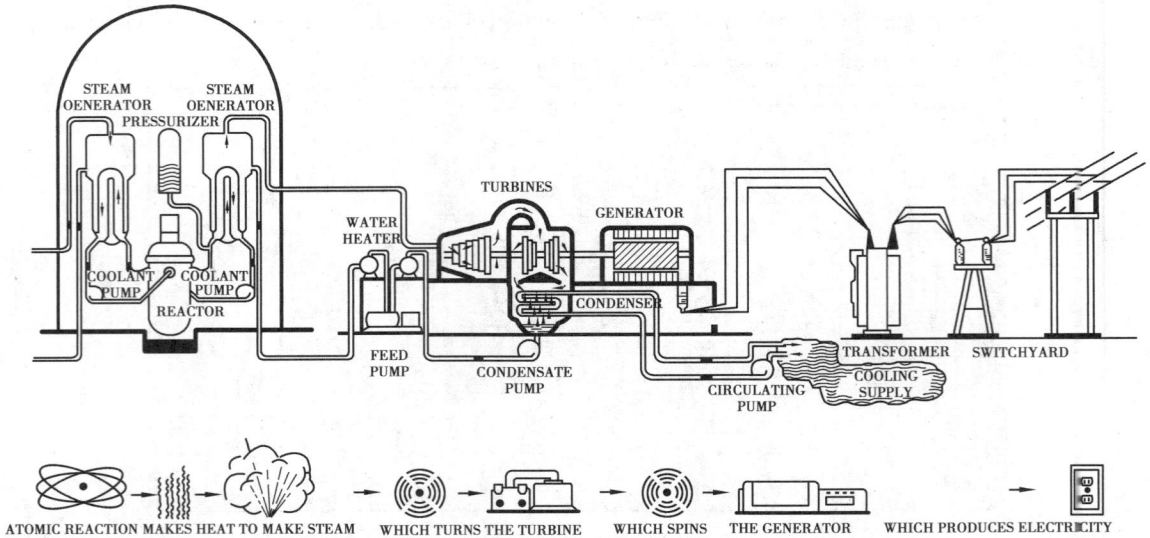

(a) Schematic diagram

(b) Exterior figure

Figure 5.8   Nuclear power plant

New energy means primary energy such as solar, wind, biomass, geothermal and ocean energy, combustible ice, etc., instead of conventional coal, hydropower, nuclear power.

It is the development and utilization of renewable energy in new and high technology. New energy is clean energy without environmental pollution, but due to its common characteristics of low energy density and high dispersion, the generation technology by new energy power is an interdisciplinary and comprehensive technology.

Solar energy from the sun's radiation energy is produced by the release of huge nuclear energy by hydrogen helium fusion in the inner hydrogen atom of the sun most of the solar energy meeting humans need comes directly or indirectly from the sun.

Although the sun's energy to the earth's atmosphere is only one percent of 2. 2 billion of its total radiant energy, it is as high as 173 000 TW, that is to say, the sun's energy per second is equivalent to 5 million T of coal.

Wind energy, hydro energy, ocean thermal energy, earth wave energy and biomass energy all come from the sun; even the earth's fossil fuels (such as coal, petroleum, natural gas, oil shale) is fundamentally the storage of solar energy since ancient times.

In the narrow sense, solar energy is limited to the direct conversion of solar radiation to photothermal, photoelectricity and photochemistry.

Figure 5.9　Solar photovoltaic power station

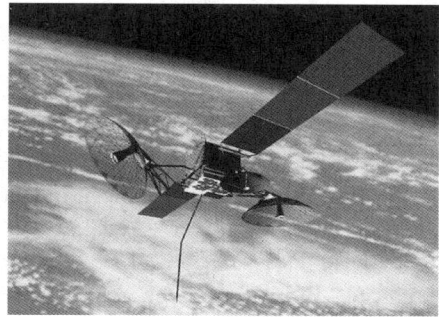

Figure 5.10　Solar panels in satellites

The generation principle of solar photovoltaic station is different from the solar thermal power. It uses solar cells to convert solar energy directly into electricity according to the photovoltaic effect.

The photovoltaic power generation system is mainly composed of electronic components, so its equipment is extremely refined, reliable, stable, long-lived and simple installation and maintenance.

Theoretically, photovoltaic power generation technology can be used for any occasion requiring power, from the spacecraft to the home power supply, as large as the MW class power station, as small as toys. In fact photovoltaic power applications exist everywhere.

5.3.5　Wind power generation technology

The wind is a natural phenomenon caused by airflow, which is created by solar radiation heat. As the sun shines on the earth's surface, the earth's surface temperature rises, and the air in the earth's surface becomes lighter and goes up by thermal expansion. With the hot air rising, cold air of low temperature inflows laterally. The rising air lands in after being cool and heavy, and then it will be heated up because of the higher surface temperature. This airflow generates the wind. From the perspective of meteorology, the wind often refers to the horizontal motion component of air, including direction and magnitude, namely wind direction and wind speed; but for flight, it also includes the vertical component which is also called the vertical or the airflow.

Figure 5.11   Wind power plant

As a huge and clean renewable energy, wind energy has attracted attention in the world in recent years. Global wind energy is about 2.74×10^9 MW, in which the available wind energy is 2×10^7 MW, 10 times greater than the total amount of water energy available on earth. Wind power has long been used to draw water and grind flour from windmills, and now people are interested in how to use the wind to generate electricity.

5.3.6   Biomass power generation technology

Biomass is an organism produced by photosynthesis. Biomass energy is one of the solar energy in the form of chemical energy stored in the biological form of energy. With the energy of the biomass as the carrier, it is directly or indirectly from the photosynthesis of plants.

In a variety of renewable energy, biomass energy is unique, since it is in storage of the solar energy, it is the only renewable carbon source that can be converted into conventional solid, liquid and gaseous fuels.

Figure 5.12   Waste power plant

Biomass which is the fourth largest source of energy after coal, oil and natural gas occupies an important position in the whole energy system. Biomass energy has always been one of the important energy sources in human life. In the world energy consumption, biomass accounts for 14% of total energy consumption, and more than 40% in the developing countries.

It is estimated that each year the fixed carbon reaches 2×10^{11} t by plant photosynthesis on earth, with the energy of 3×10^{21} J. Therefore, solar energy stored in plants, stem and leaf through photosynthesis is 10 times to the energy every year all over the world.

5.3.7   Tidal power generation technology

Tide refers to the cyclical movement of water caused by generating force produced by heavenly bodies (mainly the sun and the moon). The flow of water in the vertical direction is called tidal fluctuation, and the flow of water in the horizontal direction is called tide. Tide is a natural phenomenon in coastal areas.

Figure 5.13   Tidal power station

5.3.8   Geothermal power generation technology

The main source of geothermal energy is the heat produced by the decay of radioactive elements in the earth. Radioactive elements include uranium 238, uranium 235, thorium 232 and potassium 40. The decay of radioactive elements is the release process of atomic nuclear. Radioactive nucleus, without external force effect, can spontaneously emit photon particles of high speed as electrons and helium nuclei and form rays . Within the earth, the kinetic energy and radiant energy of these particles and rays are transformed into heat in the collision with earth. The heat of the earth is constantly released to space. This geophysical phenomenon is called earth heat flow.

Geothermal power is a new type of power generation technology in the use of underground hot water and steam as power supply. Its basic principle is similar to thermal power generation, on the basis of the principle of energy conversion. First, geothermal energy is converted to mechanical energy, and then mechanical energy is converted to electrical energy.

There are two types of geothermal power generation: geothermal steam generation and underground hot water generation.

Figure 5.14   Tibet Yangbajing Geothermal Power Plant

5.3.9   Fuel cell power generation technology

The fuel cell is a kind of power equipment that directly converts chemical energy into electrical energy. The operation theory of fuel cell is as follows: Hydrogen pole in the negative pole of the battery inputs hydrogen (the fuel pole), and inputs air or oxygen in the oxygen pole in the positive pole. The electrolyte between the positive and negative poles separates the two poles. Different kinds of fuel cells use different electrolytes, such as acidic, alkaline, molten salt or solid electrolyte. Under the effect of catalyst, fuel and oxidant in the fuel cell generate electricity and water (H_2O) through electro chemical reaction in the process of energy conversion . Therefore, the fuel cell will not produce gas as nitrogen oxides (NOX) and hydrocarbons (HC) to pollute the environment.

(1) **Basic reaction steps of fuel cell**

Transmit the reactant to the fuel cell →electrochemical reactions→ion conduction and electron conduction→product discharge .

Taking hydrogen oxygen fuel cells as an example, this reaction is the inverse process of electrolysis water.

Negative electrode: $H_2 + 2OH^-$ to $2H_2O + 2e^-$

Positive electrode: $1/2O_2 + H_2O + 2e^-$ to $2OH^-$

Battery response: $H_2 + 1/2O_2 = H_2O$

(2) **Unique advantages of fuel cells**

With the high conversion efficiency that can reach 60% , no running parts, small noise; no side effect, small fuel cells still have a similar efficiency as the larger ones. If hydrogen is used as the fuel, it can achieve "zero emission" with its discharge of water. Fuel cells are divided into solid oxide fuel cells, molten carbonate fuel cells, alkaline fuel cells, phosphoric acid fuel cells, proton exchange membrane fuel cells, etc. , according to the electrolyte it uses.

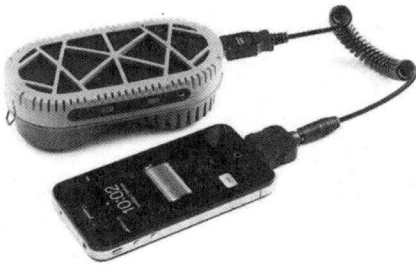

Figure 5.15　Fuel Cell Phone

Figure 5.16　Fuel Cell Vehicle

5.4　Transmission and distribution system

The transmission and distribution of electric energy is completed by the power transmission and distribution system. The power transmission and distribution system include all substations and all devices along cable or overhead conductors with different voltage levels besides the load.

(1) **Transmission**

To ensure the adequacy and reliability of power supply is the basic need of modern society.

Providing connections between power suppliers and power consumers helps reduce prices.

Power generation in other cheaper regions can be delivered to customers, and the regional systems are connected by transmission via inter-connectors to encourage cooperation for mutual benefits.

(2) **Transmission System**

Lines/transformers operating at voltages above 110 kV are usually called the transmission system.

It consists of Transmission Line and Sub-stations.

The transmission network of above 500 kV is known as UHV(Ultra-high voltage) grid.

Transmission network of 500 kV, 220 kV and 110 kV are called "Major Grid".

(3) **Distribution**

Distribution is at lower voltage levels than transmission.

Lines/transformers operating at voltages below 110 kV are usually called the distribution system.

The bulk of consumers draw power at lower voltages. This requires two different levels of distribution.

Distribution substations convert transmission voltages to distribution voltages.

Voltage classification as per IEEE 141-1993:

(1) **Low voltage (LV)**

Systems of rated voltage up to 1 000 V

Common usage: 380 V, 415 V, 480 V

（2）**Medium voltage（MV）**

Systems of nominal voltage 1 000 V and above but less than 100 kV

Common usage：4 160 V, 6 900 V, 10 kV,12 kV, 13.8 kV, 34.5 kV, 69 kV

（3）**High voltage（HV）**

Systems of nominal voltage 100 kV and above and up to 230 kV

Common usage：110 kV,115 kV, 138 kV, 220 kV,230 kV

Basis of voltage selection

System segment

Generation：MV

Transmission：HV and Extra High Voltage（220 kV +）

Sub-transmission：HV/MV

Distribution：MV/LV

Utilization：MV/LV

5.4.1 Transmission line

The transmission line is an important part of the power system, which undertakes the important task of transporting and distributing electric energy.

In the current power transmission technology, transmission lines can be divided into AC and DC transmission.

AC transmission lines are divided into high voltage lines（above 1 kV）and low voltage lines（i.e. 1 kV and below）according to the voltage level. Some are subdivided into low voltage（1 kV and below）, medium voltage（1-35 kV）, high voltage（35-330 kV）, ultra high voltage（330-1 000 kV）, ultra high voltage（1 000 kV and above）and other lines.

The DC voltage level below the nominal voltage + 800 kV is defined as high voltage, and the DC voltage level of nominal voltage + 800 kV and above is defined as UHV.

Transmission Line includes overhead lines(Figure 5.17), fully insulated shielded bus, GIL, cable（underground cables and submarine cables）.

Figure 5.17   Overhead line

Figure 5.18　Fully Insulated Shielded Bus

Figure 5.19　GIL

Figure 5.20　Power Cable

Components of Overhead Transmission Line:

Shield and Ground wire-wires primarily used for protection from lightning strikes and corresponding surges.

Insulators-insulators used to contain, separate, or support electrical conductors.

Conductors-metal cables used for carrying electric current.

Structures-structures used to support structures to hold up the conductors.

Foundation-systems use to transfer the various dead and live loads of the tower and conductors to the ground.

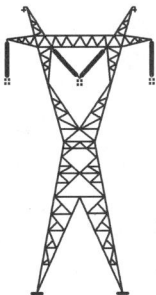

(a) Waist-Type Tower　　(b) Double Circuit Tower　　(c) Guyed-V-Tower

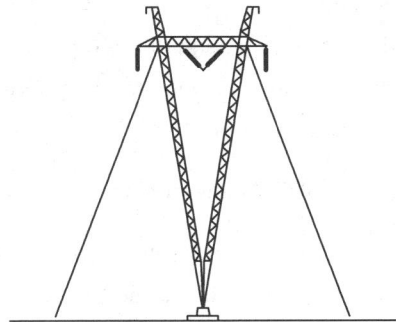

Figure 5.21　Transmission tower

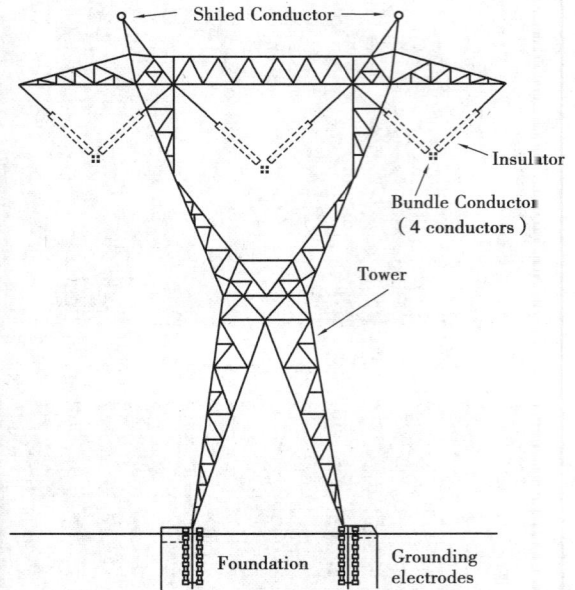

Figure 5.22   A transmission line and Structure

Classfications of overhead conductor:

Typically, aluminum or copper are used for material.

Aluminum is more common than copper for its lower cost and lighter weight. However, this comes at the price of some energy loss that does not occur with copper.

Aluminum Conductor Steel Reinforced (ACSR), including steel strands wrapped around aluminum conductors, which adds strength, is the most commonly used conductor above the ground.

Compared with the overhead line, power cable has the disadvantages of high cost, large investment and maintenance inconvenience, but it is reliable in operation, insusceptible to outside influences, beautiful, and requiring little space.

It will be the only choice especially when the field is inflammable, explosive, or filled with corrosive gases and thus is not proper to erect the overhead line.

The power cable line is one vital part of the urban power network, and its safety, stability and reliability are of great significance in the urban power network.

According to statistics, the total length of HV cable is more than 8 000 km in China, with the highest voltage reaching 500 kV.

5.4.2   AC Transmission

Increasing the voltage of the power grid can reduce the voltage drop and the wire and increase the output power to achieve the purpose of energy saving. However, the characteristics of easy rise and fall of AC voltage are suitable for high voltage transmission. Using a simple booster transformer, the alternating voltage can be promoted to thousands of volt to several hundred thousand volt. In the city, the step-down transformer will decrease by transformer voltage from tens of thousands volt to thousands of V to ensure safety. Before entering the household, the reduced voltage must meet the requirements of the municipal rated voltage or the applicable voltage power supply device.

5.4.3 HVDC Transmission

DC refers to the direct current in constant magnitude and invariant direction with time.

There are main 3 of DC transmission system: rectifier station, DC line and inverter station.

Compared with AC transmission technology, the DC transmission line has almost no reactive power loss and its transmission distance is much longer. Only two conductors are needed for each circuit. Compared with three conductors, it greatly reduces the consumption of the metal, the space of the corridor, and the investment. No alternating magnetic will be produced in the DC transmission process, and thus bring less interference to the outside (such as communications, TV and radio, etc.).

5.5 Substation

A substation is a place where voltage is changed. In order to transmit the power to the distant place, reduce the line losses, voltage step-up transformer is used to prompt the voltage into high voltage, while the step-down transformer is used to reduce the voltage where the consumers line nearby. The size of the substation mainly depends on the size of the transformer and consumers.

5.5.1 Classification of substation

A substation is a power facility that transforms voltage, accepts and distributes electrical energy, controls the direction of power flow and adjusts the voltage. It connects the power grids at all levels of the voltage through its transformers.

In the power system, the substation which is the joint of transmission and distribution, is divided into substation and power substation, according to different sizes. The one with a large scale is called substation, including voltage step-up substation and substation. The another one called as power substation generally refers to the substation with a voltage level below 110 kV.

Substations can also be divided into conventional outdoor ones (require bigger space) and GIS (gas insulated switchgear) ones.

In order to reduce the space of substation, gas-insulated metal-enclosed switcher (GIS) has been applied in recent years.

GIS puts the circuit breaker, isolating switch, bus, grounding switch, PT/CT, casing or cable terminal top separately in its seal as whole shell filled with sulfur hexafluoride gas as the insulating medium.

This kind of composite apparatus has the advantages of compact structure, small volume, light weight, no influence on the atmospheric condition, a long interval of the overhaul, no electric shock and electric noise interference.

Figure 5.23　Outdoor Substation

Its disadvantages are high price, greater demands on manufacturing and overhaul process.

(a)Outdoor　　　　　　　　　　　　　　(b)Indoor

Figure 5.24　GIS

5.5.2　Substation components

Substation includes switch equipment in the open-close circuit, bus, measurement and control instrument, instrument, relay protection device and lightning protection device, dispatching communication device, etc. Besides, some substations have reactive compensation equipment.

(1)Transformer

The transformer is the main equipment of substation. According to the form of winding, it can be divided into two winding transformer, three winding transformer and autotransformer. The latter's each phase of the high and low voltage use a common winding and a head is made out from the middle as the outlet transformer of the low winding transformer. Roughly, the voltage is proportional to the number of winding turns, and the current is inversely proportional to number of winding turns. The transformer can be divided into step-up transformer and step-down transformer according to its function. The transformer can be divided into voltage-regulating transformer with carrying load and voltage-regulating transformer without carrying load according to the switching mode of tap changer.

Figure 5. 25   Transformer

(2) **Voltage transformer and current transformer**

The voltage transformer and current transformer are used for measurement of V and I. They put the operation voltage of the high voltage equipment and bus, large current equipment and bus load or short-circuit current into lower voltage or lower current of measuring instrument, relay protection and control equipment respectively.

Under rated operating conditions, the second side voltage of the voltage transformer is 100 V, and the secondary current of the current transformer is 5 A.

The secondary side of the current transformer is often connected with the load to short circuit and can not be opened, otherwise, it will endanger the equipment and personal safety or cause the current transformer to burn down due to high voltage. Similarly, the secondary side of the voltage transformer can't be short-circuited.

(a) JDZJ-10 type voltage transformer

(b) LQJ-10 type current transformer

Figure 5. 26   PT and CT

(3) **Switch equipment**

Switching equipment is the equipment to disconnect the circuit includes circuit breaker, isolation switch, load switch, high voltage fuse, etc.

1）Circuit breaker. A circuit breaker is used to connect and disconnect the circuit under the condition of normal operation of power system. It also can disconnect the fault equipment and circuitry automatically under the control of relay protection device in a failure. It also has an automatic reclosing function. In China, air circuit breakers and hexafluoride circuit breakers are more commonly used in the +220 kV substations.

Figure 5.27　CB

2）Disconnecting switch. The main function of the disconnecting switch is to isolate the voltage when the equipment or circuit is overhauled, so that the circuit can form an obvious disconnection point to ensure the safety of the maintenance personnel. It cannot disconnect the load current and the short-circuit current, like the circuit breaker. In the case of power failure, the circuit breaker should be pulled off before disconnecting switch pulled off to show the open circuit opprently. On contrary, to connect/recover the transimission peration, the disconnecting switch should be closed before the circuit breaker is closed. An incorrect operation procedure will cause equipment damage, personal injury and other accidents.

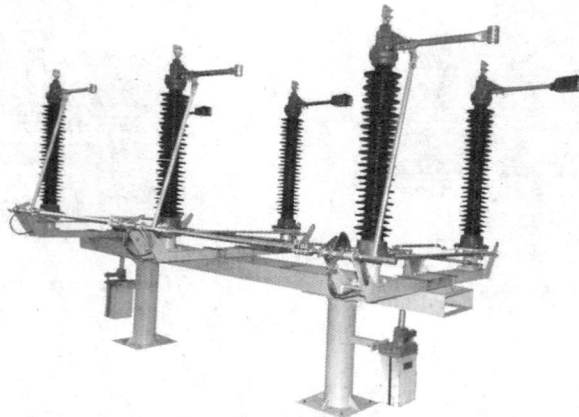

Figure 5.28　Isolation switch

3) Load switch. A load switch is used to disconnect the load current during normal operation, but it has no ability to disconnect the fault current. Generally, it is used with high voltage fuse for transformer of +10 kV voltage and not often used the outlet.

(4) Lightning protection equipment

Lightning protection equipment installed in substation includes main lightning rod and lightning arrester. Lightning rod leads lightning current into the earth to prevent the substation from the direct lightning strikes. When lightning in the vicinity of a substation falls on the line, the lightning wave will enter the substation along the wire to produce overvoltage. In addition, the circuit breaker operation will also cause overvoltage. Arrester is used when the overvoltage exceeds a certain limit to automatically discharge to the ground, reduces the discharge of voltage protection device, and then automatically extinguish quickly to ensure the normal operation of the system. At present, the most common one is zinc oxide lightning arrester.

Figure 5.29 Zinc Oxide arrester

5.6 Power supply and utility technology

Electricity energy cannot be stored. Electricity generated capacity is based on demand. Total power drawn by consumers fluctuates depending on the time of day and seasons.

There are three types power load: load(100% of the time), Intermediate loads(fed <100% of the time), Peak load(May occur 0.1% of the time).

5.6.1 Power load

(1) Basic concepts of power load

Power load, also known as load, has two definitions: one is equipment which only consumes electrical energy and users, such as important load, general load, dynamic load, lighting load, etc. The other one is the power or current size consumed by electric equipment, such as light load (light load), heavy load (heavy load), empty load (no-load), full load (full load). The specific defini-

Typical Daily Weekday System Load Profile

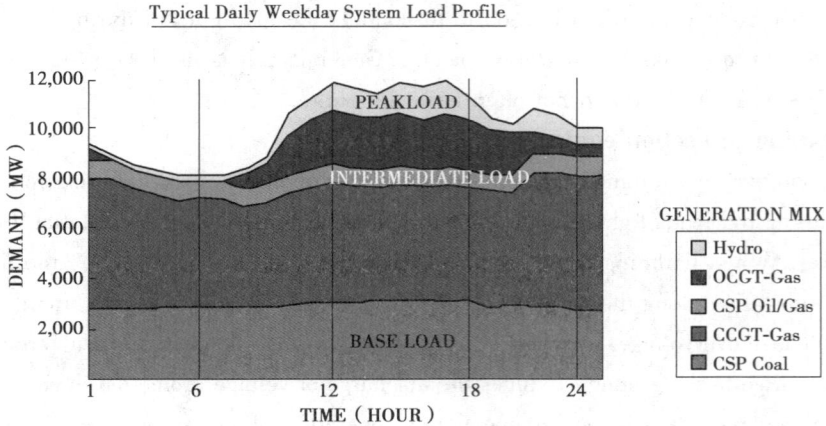

Figure 5.30   Demand Curve

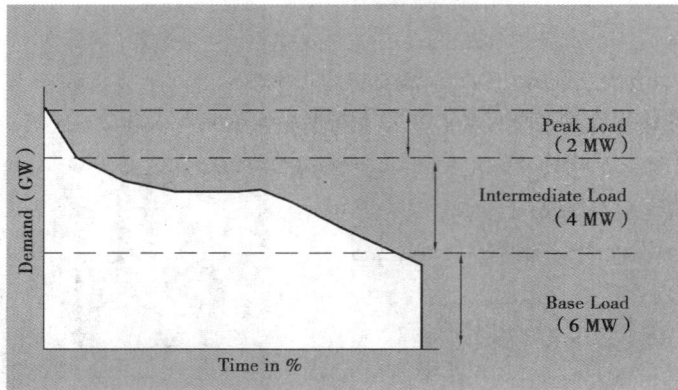

Figure 5.31   Load Duration Curve

tion of power load depends on the specific circumstances. Power load includes asynchronous motor, synchronous motor, all kinds of electric arc furnace, rectifying device, electrolytic equipment, refrigeration and heating equipment, electronic equipment and lighting facilities. They belong to industry, agriculture, business, transportation, scientific research institutions, cultural entertainment and all kinds of power users. According to the different load characteristics of power users, the power load can be divided into various industrial loads, agricultural load, transportation load and people's living power load.

(2) Grade of power load

The power load shall be classified according to the requirement level of the power supply on reliability and the degree of loss or influence caused by the interruption of power in the politics and economy. They are as following provisions:

① First class load. In accordance with the following circumstances, it should be regarded as the first level load. The first class load should be powered by two independent power supply, and emergency power supply should be added for the very important load.

(a) Interruption of power supply will cause personal injury and death.

(b) Interruption of power supply will cause political and economic losses. For example, major

88

equipment damage, major product scrap, a large number of scrapped products produced by important raw materials, a disrupted continuous production process of key enterprises in the national economy which needs a long time to recover.

(c) Interruption of power supply will affect the normal operation of the electricity-consuming units with significant political and economic significance. For example, important transportation hub, important communication hub, important hotels, large stadiums, a large number of public places, which are often used for international activities where crowds of people gather, and the important power load in the power unit. The first class load, when the interruption of power supply produces such situations of load as poisoning, explosion, fire and others as well as the important places where the power supply is not allowed to interrupt the load, should be considered as a particularly important load.

② Second level load. The second level load requires two circuits and two transformers.

(a) Interruption of power supply will result in great political and economic losses. For example, the main equipment is damaged, a large number of products will be scrapped, the continuous production process will be disrupted for a long time to recover, and the major enterprises will reduce production.

(b) Interruption of power supply will affect the normal work of important electrical units. For example, transportation hub, an important power load in power units such as communication hub, as well as the interruption of power supply, will cause disorders of important public places, such as large cinema and large shopping malls.

③ Third level load. Loads not belonging to first level load and second level load are third level load. The third level load has no special requirement for power supply.

5.6.2　Power quality (PQ)

Power quality is the quality of electricity in the power system. The ideal electricity should be a perfectly symmetrical sine wave. However, some factors may cause the waveform to deviate from the symmetric sinusoid, thus generating the power quality problem.

Power quality problems can lead to electrical equipment or system failure because of voltage, current, or frequency deviation.

The assessment parameters of power quality include the frequency deviation, voltage deviation, voltage fluctuation and flicker, three-phase imbalance, transient overvoltage, waveform distortion (harmonic), voltage sag, temporary interruption, and power supply continuity problem.

Strictly speaking, the main indicators of PQ are voltage, frequency and waveform.

Generally speaking, it refers to high quality power supply, including voltage quality, current quality, high quality power supply and consumption.

5.6.3　Electricity market

The electricity market refers to organization structure, operation management and operation rules of electric power industry based on market economy principle.

The electricity market is also a concrete execution system, including trading place, transaction management system, measurement and settlement system, information and communication system, etc.

5.6.4 Power operation and control

The operation mode of power system is the technical scheme of power system production and operation compiled by power system dispatching department. From the time, it is divided into year, month and day; from the use, it is divided into normal, maintenance, after the accident, etc. They permeate each other and connect with each other closely.

① The annual operation mode of power system.

② The normal operation mode of power system.

③ The operation mode of power system maintenance.

④ The operation mode after power system accident.

5.6.5 Power safety

(1) The damage to the human body

Electrical damage to the human body mainly includes electric shock and electrocution.

1) Electric shock. Electric shock is the damage caused by the electrical current passing through the body, destroying the normal work of the heart, nervous system and lungs. It can arouse muscle twitch, internal tissue injury, cause febrile numbness, nerve palsy, which leads to coma, suffocation, no heart beating and even death. Most cases of electrocution are caused by shock. When a person touches a charged wire, a leaking shell or other charged devices and a lightning strike or capacitor discharge, it may cause an electric shock. The size of the current is the main cause of shock damage, and the safety current of the human body is 30 mA. The human body's tolerance to electricity is related for the following factors:

① The type and frequency of the current.

② The magnitude and duration period of current.

③ The current path through the human body .

④ The magnitude of voltage.

⑤ The physical condition of the person.

2) Electrical injury

Electric injury refers to the local effect by the thermal effect, chemical effect and mechanical effect of the current on the human body. It can be caused by electric current directly through the human body, or by arc or electric spark. Damage to electrical injury includes arc burns, electric brand, skin metallization, electrical mechanical injury, electro-optical eye forms (fracture or injuries caused by electrician falling down in the overhead working accidentally are also electrical injuries), with clinical manifestations of dizziness, heartbeat, sweating or nausea, vomiting, and skin burn pain.

(2) Safety voltage

When the body resistance is certain, the higher the voltage of the body touch is, the greater the current will get through the body, the more serious damage will bring to the human body. But it is not true that every time people touch the power supply, it will cause harmness to the human body.

In daily life, when we touch the poles of ordinary dry cells, the body doesn't feel anything, because the voltage of the ordinary dry cell is as low as DC 1.5 V. When the voltage on the human body is below a certain value, the voltage will not cause serious injury to person in a short period of time, and we call this voltage as the safety voltage.

In order to determine the safety conditions, Normally, the safety voltage is used to estimate safety or not.

Generally speaking, in a dry and less dangerous environment, safe voltage is 36 V; in a humid and dangerous environment (such as metal container, pipe welding repair), safe voltage is 12 V. When the electric current gets through the body, the voltage restrict the current to a small range, and personal safety can be ensured in a certain extent.

(3) The way of electric shock

① Single phase electric shock. Electric shock single phase lines and ground.

② Two phase electric shock. Electric shock between two phase lines.

③ Contact electric shock. Electric shock caused by touching the conductor.

④ Step electric shock. As shown in the diagram, a circular electric field is generated near the ground where the wire contacts. That is, the voltage distribution of the radial voltage decreases from the contact point to the periphery with the contact point of the high voltage line as the center of the circle. The potential at the center of the circle is equal to the potential on the high voltage wire. The farther distance from the center is, the smaller the voltage is.

The potential difference between the two feet forms a loop through the body, causing an electric shock accident. The closer distance from the wire's landing point is, the greater the step voltage is, the more dangerous to the person is.

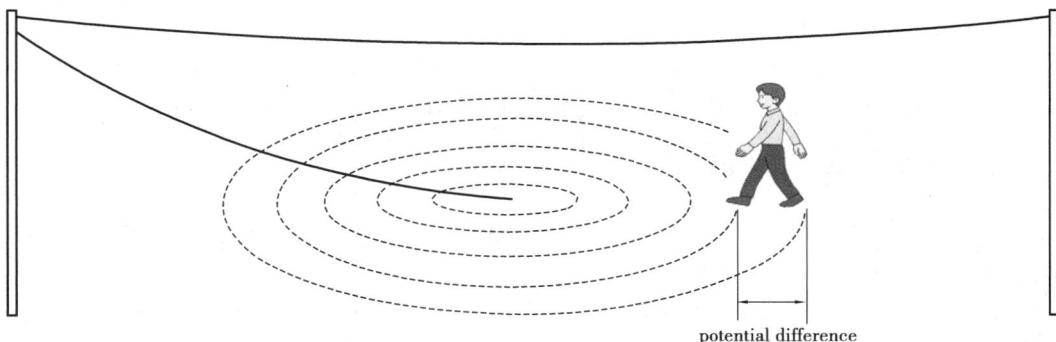

potential difference

Figure 5.32   Step electric shock

(4) Never touch the victim suffering from electric shock

Specific rescue measures are generally pulling, cutting, picking, dragging and padding.

① Pulling. Disconnect the power switch and the power supply.

② Cutting. Cut off the power supply by using electric pliers, spades, picks, knives, axes and other sharp tools with reliable insulating handle. When cutting, attention should be paid to prevent the live wires from falling off and touching people around them.

③ Picking. If a wire falls on the electric shock victim or pressed under the body, use a dry stick, bamboo to break the lead.

④ Dragging. It refers to the use of insulation articles in the hands to drag and drop electric shock from the power source.

⑤ Matting. If an electric shock victim's finger is gripped by a wire due to cramps, a dry board or rubber insulation pad can be put into the body first to make it insulated from the earth, and the power supply should be cut off.

5.7　Protection and remote control system

5.7.1　Protection system

In power system operation, the external factors (such as lightning, birds, etc.), internal factors (insulation aging, damage, etc.) and also improper operation, can cause all kinds of faults and abnormal states. Common faults include single-phase grounding, three-phase short circuit, two-phase short circuit, two-phase grounding and short circuit, broken line, etc.

The abnormal operating states include overcharged, overvoltage, non-full phase operation, oscillation, subsynchronous resonance, synchronous generator short-time loss of magnetic asynchronous operation, etc.

Relay protection and safety automatic device are used for rapid removal of fault and elimination of the abnormal condition when faults and abnormal operation occurs. When the power system fails or is in dangers, it can send a warning signal in time or issue the tripping order directly to terminate the event.

The relay protection procedure is to operate from circuit breaker automatically, quickly and selectivity and then reflect the abnormal running state of electrical components.

5.7.2　Telecontrol system

Telecontrol system for power system is a remote monitoring and control technology for dispatching service of power system.

It can manage and monitor the operating conditions of many factories, plants, stations, equipment and components which are widely distributed in various regions.

Due to the characteristics of energy production, energy center and load center are always far from each other. Power system distributes in a wide area, where power plant, substation and power dispatching center are in dozens, hundreds or even thousands of km from the users.

To manage and monitor the operation condition of large factories, plants, equipments and components of wide distribution, a technical means (the remote control technology) must be relied on, and it is impossible to use the usual mechanical/electrical contact to transmit control information or feedback data.

It can convert operating conditions and parameters in all plants and stations to signal forms of easy transmission, These forms can be modulated transmitted to the dispatch department by the information transmission channel.

In the dispatch department, the signals are demodulated to the original signals for the monitoring and use of the dispatcher.

Some orders of dispatchers can also be transmitted to the remote plants and stations through a similar process to drive the controlled objects.

This process involves telemetry, remote communication, remote regulating and remote control, therefore, the telecontrol technology is the combination of four remote controls.

With the rapid development of electronics technology, computer technology and communication technology, the power telecontrol system has undergone significant technical renovation.

Chapter 6

High-voltage and Insulation Technology

6. 1   The source of high-voltage disciplines

6.1.1   Generation and transmission of electric energy

① Early electric power systems

The first public power station was put into service in 1882 in London. It contained three features: to produce direct current at low voltage with a service limited to highly localized areas, to be used mainly for electric lighting.

The first major AC power station was commissioned in 1890 at Deptford. It contained two features: to supply power over a distance of 28 miles, to produce alternative current at a high voltage of 10 kV.

② Two principle factors influence the development of power transmission networks

To make full use of economic generation, transmission networks must be interconnected for pooling of generation in an integrated system.

Bulk transfer over long distances.

③ Power transfer capability

Power transfer capability P of AC system is approximately considered as:

$$P = U^2/Z \tag{6.1}$$

where Z is the wave impedance of a transmission line, U is the operating voltage.

Table 6.1   Power transfer capability of AC system and its relationships with operating voltage and transfer distance

$U/$ kV	220	330	500	750	1 000
Z/Ω	400	303	278	256	250
$P/$MW	121	360	900	2 200	4 000
Transfer distance/km	100-300	200-600	1 000 around	>1 000	>1 000

Power transfer capability P of a bipolar DC system is approximately considered as:

$$P = 2U_{d}I_{d} \qquad (6.2)$$

where U_d is the operating voltage of one line and I_d is the current flowing through the line.

Table 6.2　Power transfer capability of bipolar DC system and its relationships with operating voltage

U/kV	220	330	500	750	1 000
Z/Ω	400	303	278	256	250
P/MW	121	360	900	2 200	4 000
Transfer distance/km	100-300	200-600	1 000 around	>1 000	>1 000

④ Development of power systems in different voltage

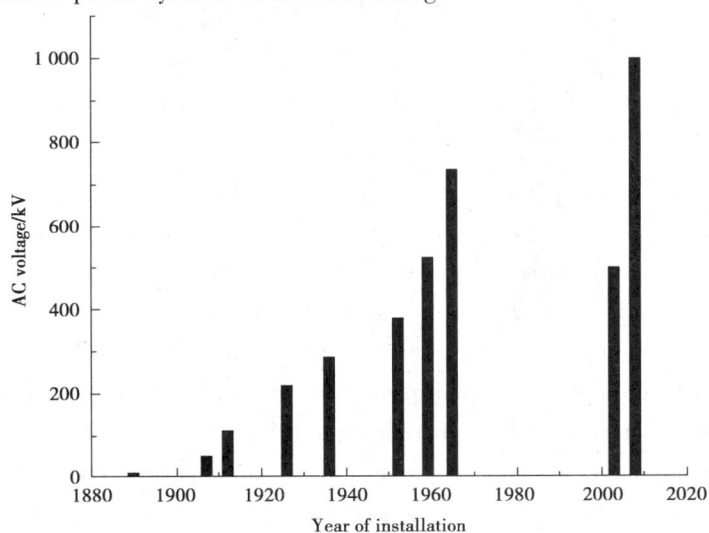

Figure 6.1　Major AC systems in chronological order of their installations

1890	10 kV	Deptford	1952	380 kV	Harspränget-Hallsberg
1907	50 kV	Stadtwerke München	1959	525 kV	USSR
1912	110 kV	Lauchhammer-Riesa	1965	735 kV	Manicouagan-Montreal
1926	220 kV	N. Pennsylvania	2003	500 kV	Three Gorges (China)
1936	287 kV	Boulder Dam	2008	1 000 kV	Jindongnan-Nanyang-Jingmen (China)

6.1.2　Voltage class

Definition of High Voltage: The IEC and its national counterparts (IET, IEEE, VDE, etc.) define *high voltage* circuits as those with more than 1 000 V for AC and at least 1 500 V for DC. In electric power transmission engineering, *high voltage* is usually considered as any voltage over approximately 35 kV.

　　AC systems:　　High voltage levels: 35-220 kV

　　　　　　　　　　Extra-high voltage (EHV) levels: ⩾330 kV and <1 000 kV

　　　　　　　　　　Ultra-high voltage (UHV) levels: 1 000 kV and above

DC systems： HVDC levels：600 kV and below

UHVDC levels：above 600 kV

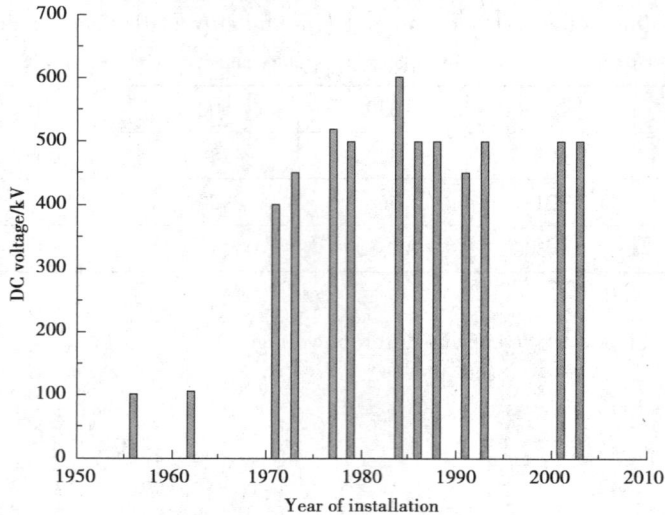

Figure 6.2　Major DC systems in chronological order of their installations

HVDC permits a higher power density as compared to AC transmission. HVDC provides an economic solution for interconnecting asynchronous AC systems (*back-to-back* installation).

Description of operating voltage：normal operating voltage does not severely stress the power system's insulation. Only in special circumstances, for example under pollution conditions, operating voltages may cause problems to external insulation. Operating voltage determines the dimensions of the insulation which forms part of electrical equipment.

Voltage stresses on power systems arise from various over voltages. Over voltages include external over voltage and internal over voltage. External over voltage is associating with lightning strokes on lines. Internal over voltage is generated by changes in the operating conditions of systems, such as switching operation (switching overvoltage) and faults on systems or fluctuations in loads (voltage rise and fluctuation overvoltage).

6.1.3　Testing voltages

It is necessary to test HV equipment during its development stage and prior to commissioning. The magnitude and type of test voltage vary with the rated voltage of a particular apparatus.

Basic classification of testing voltages is as follows：

Testing with power frequency voltages

Testing with lightning impulse voltages

Testing with switching impulse voltages

Testing with DC voltages

Testing with very low-frequency voltages

6.2　**Main contents**

High voltage engineering consists of generation, measurement, control of high voltages, dielectric discharges and electrical insulation breakdown, overvoltage and their protection, and electrical insulation condition monitoring and diagnosis, etc. The course includes three main parts: dielectrics & electrical insulation, overvoltage & insulation coordination in electric power systems, and high voltage testing techniques. What are their relationships? Reliable insulation is related to the application of high voltage systems. Electrical insulation is the physical basis of high voltage engineering. High voltage tests ensure a high voltage system to be operated safely. May over voltages have much greater magnitudes than normal operating voltage? They do severely stress the power system's insulation.

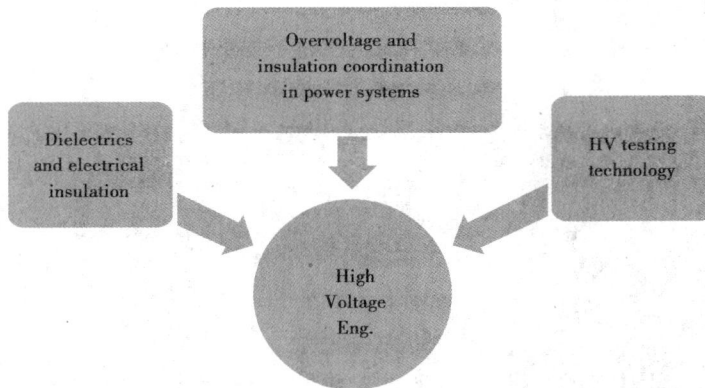

Figure 6.3　Main contents

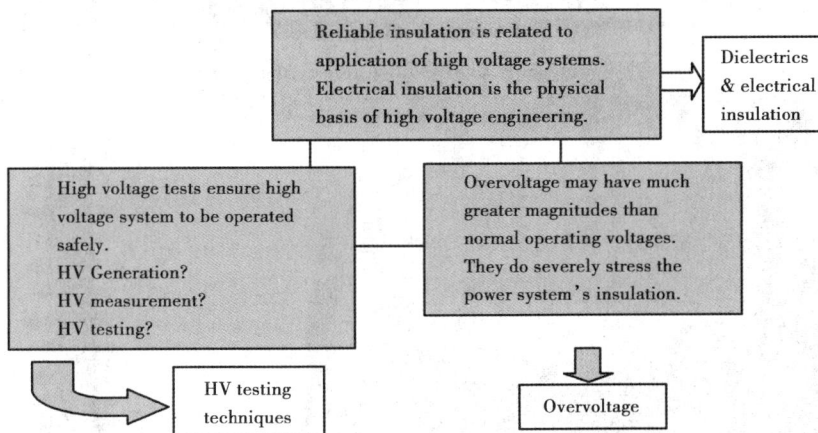

Figure 6.4　Relationships of three main parts

Let's take a look at the picture of the HV equipment.

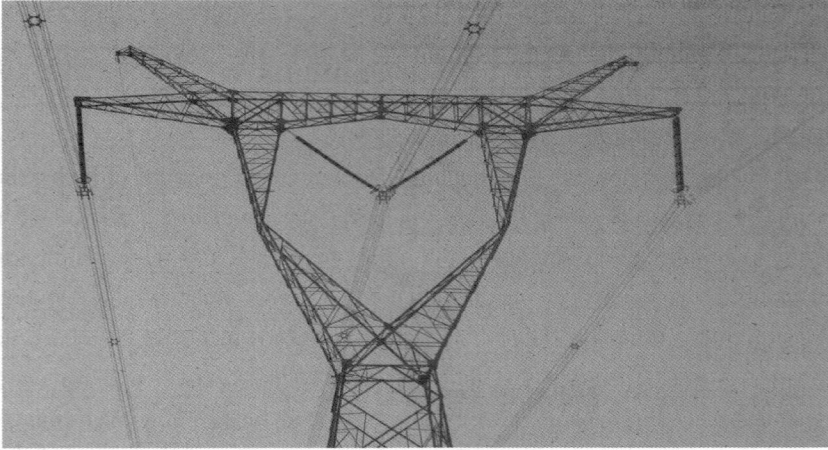

Figure 6. 5   AC 750 kV transmission line

Figure 6. 6   Oil-filled transformer

Figure 6. 7   Metal oxide arresters

Figure 6.8   Testing transformer

Figure 6.9   Impulse voltage generator

The contents of electrics and electrical insulation mainly involve in dielectrics and insulating material, insulation structures and electric field distribution, and voltage applied on electric insulation. There are many important things to learn: properties and phenomenon of dielectrics in high electromagnetic fields, how to design proper insulation structures to satisfy requirements, and the consideration of AC, DC, impulse, and combined voltages for insulation structure designing.

The contents of high voltage testing techniques are to answer three questions: How to generate high voltage? How to execute high voltage experiment? How to measure high voltage? Therefore, the following points should be paid attention to : economic and flexible HV testing equipment, HV testing programs and standards, measurement of High field, tiny quantity of parameters of HV equipment of systems, transient parameter measurement.

6.3　Dielectric insulation

Dielectric can be divided into gas dielectric, liquid dielectric and solid dielectric. The actual insulation structure usually uses combination of several dielectric insulations. The electrical strength of a dielectric is limited. Beyond the limit, the insulation material or dielectric will be destroyed. It is more convenient to determine the insulation configuration of power line and equipment by studying the electrical characteristics of dielectrics.

Gas is the most widely used insulating medium in power system. Gas dielectric, especially air, is the main insulating medium in the power system. The external insulation of transmission lines and electrical equipment is air.

When the electric field strength of the gas gap reaches a critical value, the electric current in the gas gap would increase rapidly, and the gas medium will lose its insulation ability and be broken down. This phenomenon is called the breakdown of the gas medium, also known as gas discharge. It has different discharge forms: when the pressure is low and the power supply is small, a glow discharge is filled with gaps; it would be a spark discharge or an arc discharge with a higher power supply. In an extremely inhomogeneous electric field, corona discharge will be generated at the strongest local electric field. Under the action of electric field, the generation and disappearance of point particles in the gas gap determine the strength and development of discharge phenomena in the gas.

The liquid dielectric has the same fluidity as the gas dielectric which is self-healing after the breakdown, but its electrical strength is higher than that of the gas. High voltage electrical equipment made with liquid medium instead of gas medium is small and material saving. However, most of the liquid medias are flammable and prone to oxidation and deterioration, resulting in water, gas, acid, oil sludge, etc., leading to the deterioration of electrical properties. Liquid dielectric, also known as insulating oil, is liquid at room temperature and ACTS as insulation, heat transfer, immersion and filling in electrical equipment. It is mainly used in electrical equipment, such as transformer, oil circuit breaker, capacitor and cable. The insulating oil in the circuit breaker and capacitor also has the function of arc suppression and energy storage respectively. There are three kinds of liquid dielectric: mineral insulating oil, synthetic insulating oil and vegetable oil. Mixed oils are often used to improve some properties.

Solid dielectric is widely used as the inner insulation of electrical equipment, and common ones include insulating paper, cardboard, mica, plastics, etc., while the solid dielectric used for making insulators includes porcelain, glass, silicon rubber, etc.

6.4　Measurement of high voltages

6.4.1　Classification of insulation tests

High voltage and insulation technology is a subject based on experimental research and closely combined with the experiments, due to the dielectric theory is not perfect enough. Therefore, many problems of high voltage and electrical insulation must be explained through experiments; insulation design and fault diagnosis of electrical equipment must also be completed through experiments. The same equipment is often subject to multiple tests before making a relatively accurate judgment and conclusion. Potential defects can be identified by conducting insulation tests, measuring changes in these characteristics, and then measures can be taken to eliminate hazards. The purpose of electrical equipment insulation test is to judge whether the equipment can be put into operation, whether there are defects or not, to prevent equipment damage, and to ensure the safe operation of the equipment. There are two kinds of insulation tests: destructive test and non-destructive test.

As a non-destructive test method, electrical insulation preventive experiment has become one of the important measures for the safe and reliable operation of modern power system. Common insulation preventive test items include measurement of insulation's resistance and absorption ratio, measurement of current's leakage, measurement of loss angle tangent of dielectric, and measurement of insulation's oil performance.

The insulation of electrical equipment is not only affected by the working voltage for a long time, but also affected by various over-voltage. In order to test the insulation strength of electrical equipment, it is necessary to carry out various high-voltage withstand tests when the equipment is delivered from the factory, installed and commissioned or overhauled. The high-voltage withstand tests is destructive test method, which includes power frequency withstand voltage test, DC withstand voltage test, and impulse withstand test.

6.4.2 Power frequency withstand test

High voltage test transformer is one of the most basic and indispensable main equipment in high voltage laboratory. It is used as a power supply and is a component of AC, DC and impulse voltage test equipment. In high-voltage laboratory, the high-voltage test transformer or its cascade device is usually used to generate the high-voltage test voltage. However, for the tested samples with large capacitance, such as cables and capacitors, the power-frequency high-voltage can be obtained by using the cascade resonance circuit.

At present, the most commonly used measurement methods for testing products are measurement of the peak value of ac voltage via ball gap or peak voltage meter and measurement of the effective value of ac voltage via electrostatic voltmeter (peak voltage meter and electrostatic voltmeter are often used in conjunction with the voltage divider to expand the range of the meter); In order to observe the waveform of the voltage, the output signal can also be obtained by the oscilloscope from the low-voltage side of the voltage divider. Voltage transformers with low voltage instruments are usually used in power system to measure high voltage. Due to the high dispersion of high voltage discharge, the demand on measurement accuracy is not high. According to the current national and international standards (IEC), the margin of error is required to be no more than ±3%. Figure 6.10 is the schematic diagram of the measuring methods of several kinds of power frequency high voltage withstand test. It is the most convenient method to get the output voltage value of the high voltage side by multiplying the voltage value measured from the primary side (P_1 , P_2) of the transformer by the voltage ratio, but the error of this method is usually large and often plays the role of auxiliary indication.

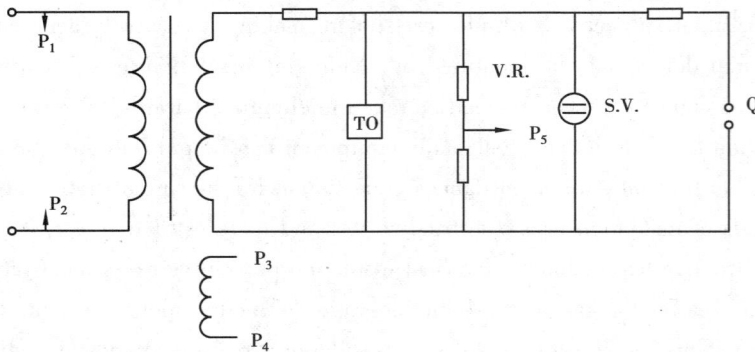

Figure 6.10   Schematic diagram of the measuring method of power frequency high voltage

It is the most effective and direct method to test the insulation strength of electrical equipment. It can be used to determine the level of the insulation tolerance voltage of the electrical equipment and to judge whether the electrical equipment can continue to operate. In the power frequency withstand voltage test, a much higher test voltage is applied to the insulation material of the electrical equipment than the working voltage, which reflects the insulation level of the electrical equipment. The with stand voltage test can effectively detect the various defects that lead to the reduction of the electrical strength of the insulation material. In order to avoid damage to the equipment during the test, the power frequency withstand test must be carried out after a series of non-destructive tests. Only after passing the non-destructive test can the power frequency withstand test be allowed.

The basic schematic diagram of the power frequency high-voltage test (including the withstand test) via the test transformer or its cascade device is shown in Figure 6.11. Since the output voltage of the test transformer must be uniformly adjusted within a large range, its low-voltage winding should be powered by a voltage regulator. The voltage regulator should be able to adjust the voltage continuously and steadily to satisfy the prescribed voltage boosting speed, so that the voltage on the high-pressure side changes within the range of 0-U (test voltage).

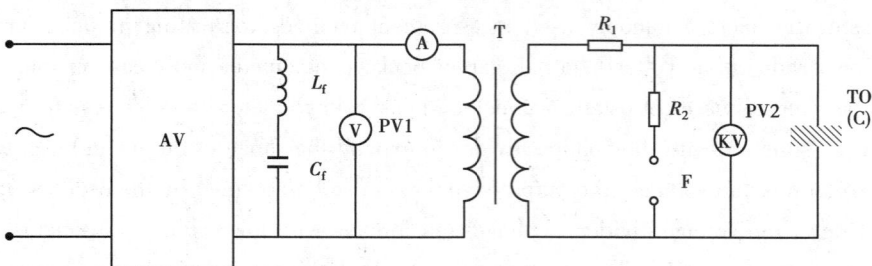

Figure 6.11   Schematic diagram of power frequency high-voltage test

The implementation method of the power frequency withstand voltage test is as follows: the voltage acting on the tested product is raised according to the specified voltage boosting speed until it is equal to the required experimental voltage U, at the calculation time is started. In order to make the

102

insulation of the defective test products develop partial discharge or a complete breakdown in time, it should be kept for a period of time after reaching U, generally taking 1 min. If no insulation breakdown or local damage is found during this period (it can be judged by abnormal phenomena such as sound, decomposition of gas, smoke, sharp swing of the pointer of the voltmeter and sharp increase of the indicator of the ammeter), it can be considered that the power frequency withstand test of the test product has passed.

6.4.3　DC withstand voltage test

Electrical equipment often requires insulation tests under DC voltage. Some large-capacity AC equipment, such as power cables, also often use DC withstand voltage test to replace AC withstand voltage test. HVDC power equipment must be subjected to the DC high voltage test. In addition, with the development of HVDC technology, more and more HVDC transmission projects have emerged, so it is necessary to carry out a variety of HVDC high voltage tests also use HVDC as a power source.

It has a great practical significance to test the insulation of electrical equipment via DC high voltage. DC high voltage test is to test the withstand strength of electrical equipment. It can reflect the problems of damp, deterioration and local defects of equipment. At present, DC withstand voltage test is widely used in the insulation preventive test of generator, motor, cable and capacitor. Compared with AC withstand voltage test, DC withstand voltage test of insulation has some characteristics, as followed:

① There is no capacitive current under DC, which requires that the power supply should be with capacity. In addition, HVDC can be produced by cascade method, so the experimental equipment can be produced lightly, which is suitable for the requirements of field preventive tests. For the test of equipment with large capacity, it is not easy to capture the purpose via AC withstand voltage test. However, it only needs to determine insulation leakage current (up to milliamp level) via DC withstand voltage test. Therefore, the size of DC test equipment could be small enough and relatively portable, which is more suitable for the requirements of field preventive tests.

② The leakage current can be measured at the same time during the DC withstand voltage test. The voltage-current curve can show the concentration defect or damp of the insulation material effectively, which provide supplementary information about the insulation state.

③ Compared to AC withstand voltage test, DC withstand voltage test can detect the insulation defects near electrodes more easily. The reason is that there is no capacitive current flowing through insulation under DC voltage.

④ Under DC high voltage, partial discharge is weak and does not accelerate the decomposition of organic insulation materials aging. The DC high voltage has the feature of non-destructive testing to a certain extent. However, compared with AC withstand voltage test, DC withstand voltage test has the disadvantage that DC withstand voltage test is not as close to the actual situation of equipment. Figure 6.12 shows the schematic diagram of DC high voltage test wiring.

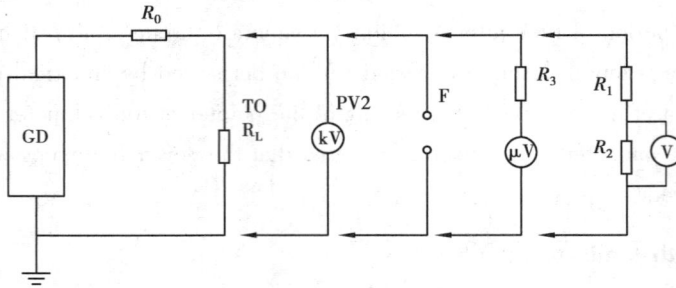

Figure 6.12   Schematic diagram of DC high voltage test wiring

In general, the current of DC high-voltage testing is small in the range of several milliamps to tens of milliamps, but some test samples of critical leakage current are quite large before homeopathic breakdown. For example, the leakage current of long air gap will reach the ampere level immediately before breakdown. Such a large leakage current will cause a large voltage drop on the device and make the measurement inaccurate. So the DC high voltage test should be based on different test samples and requirements to choose the appropriate power supply capacity.

Another important problem in DC high voltage test is: In order to prevent the discharging of the test samples or the short circuit between the generator output and the ground, a protection resistance R' needs to be threaded between the test and the high voltage output.

In addition, the selection of DC withstand voltage is also an important issue. Due to the dielectric loss of insulation materials under DC is very small, and the development of partial discharge is far weaker than that under AC, so the electrical strength of insulation materials under DC is generally higher than that under AC. This situation must be considered when choosing the DC test voltage value. The voltage of DC withstand voltage test usually uses a higher, and it is generally more than twice the rated voltage. Moreover, once abnormal phenomena appear, the DC test should be stopped in time. For example, for the generator stator windings, the voltage of DC withstand test are at 2-3 times rated voltage; for oil paper insulated power cable, the DC withstand test voltage of 2 to 10 kV cable takes 5 times the rated voltage, the one of 15 to 30 kV cable takes 4 times the rated voltage, the one of 15 to 30 kV cable takes 4 times the rated voltage, the one of 35 kV and above cable takes 2.6 times the rated voltage. The time of DC withstand test is longer than that of the AC withstand test. Therefore, DC withstand test of the generator in each stage is 0.5 times the rated voltage rise in stages, each stage stays 1 minute to read the leakage current value.

In addition, the HVDC device is also used to carry out various HVDC tests on HVDC transmission equipment, Such as the DC breakdown characteristics of various typical air gaps, the DC corona and its derivative effects on EHV DC transmission lines, the electrical performance of various insulation materials and insulation structures under HVDC, and the DC withstand voltage test of various HVDC transmission equipment. It should be pointed out that the negative polarity test voltage is usually used in general DC high voltage test as lightning impulse withstand voltage.

6.4.4 Impact withstand voltage test

In addition to withstanding long-term operating voltage, the electrical equipment may also withstand short-term lightning overvoltage and switching overvoltage during operation. In order to study the insulation performance of electrical equipments under lightning overvoltage and switching overvoltage, many electrical equipment need to carry out the impulse voltage test after type test, factory test or overhaul. The impulse voltage generator is a device which produces impulse voltage wave, and it is also one of the basic equipment in high voltage laboratory. With the continuous improvement of the transmission voltage, the voltage generated by the impulse voltage generator must be increased correspondingly to meet the test requirements. The nominal voltage of the largest impulse voltage generator in the world has reached 6 000 kV.

The lightning impulse withstand voltage test of insulation in electrical equipment adopts a three-time impulse method, that is, applying three times positive and three times negative lightning impulse test voltage ($1.2/50$ μ full wave), respectively. For the internal insulation of transformer and reactor equipment, additional lightning impulse cut-off wave ($1.2/2$-5 μs) withstand voltage test should be carried out.

In the full-wave withstand voltage test of internal insulation shock, a spherical gap should be connected in parallel with the test sample, and discharge voltage of spherical gap should be 15%-20% higher than the test voltage (transformer and reactor test sample) or 5%-10% (other test samples). In the process of wave modulation of the impulse voltage generator, sometimes there would be unintentionally high impulse voltage, which could result in unnecessary damage of the tested sample. The parallel ball gap can play a protective role under the condition.

One of the challenges in the impact withstand voltage test of internal insulation is how to detect local damage or faults in the insulation material. This reason is that the action time of impulse voltage is quite short, and sometimes the incoherent local damage is left behind in insulation materials. For example, there is often no obvious difference between the defects of winding and the ones of pie-to-wire in power transformers. At present, the most commonly used detection method is to take the current oscillogram at the neutral point of the transformer and compare the obtained oscillogram with the typical one taken in the intact transformer of the same type, the oscillogram taken in the presence of man-made faults. On the based, it is often not only to judge the occurrence of defects but also can determine their location, which greatly simplifies the work of diagnose and location of the defects of power transformers.

The impulse voltage test of external insulation of electrical equipment adopts a fifteen-time impulse method, which is, 15 positive and 15 negative shocks are applied to the tested products, and the interval time between the two shocks should not be less than 1 minute. If the number of flashover or breakdown does not exceed two times, the external insulation test can be considered qualified. The method of operation for internal and external insulation is exactly the same as that of lightning impulse full wave test.

6.5 Overvoltage protection and insulation coordination

Overvoltage and insulation cooperation is one of the important parts of the power system, including the generation mechanism and restriction measures of lightning overvoltage and operating overvoltage, and the determined principle and method of insulation cooperation by the characteristics of different overvoltage.

6.5.1 Lightning protection of transmission lines

The transmission line is located in the wilderness and vulnerable to the lightning strike. The tripping accident caused by the lightning strike lines accounts for a large proportion of the total accidents in the power grid. At the same time, the lightning wave invading the substation when lightning strikes the line is also the main factor threatening the substation. Therefore, the lightning protection of the line should be paid more attention.

There are two kinds of atmospheric overvoltage on transmission lines, one is caused by direct lightning strike on the line, namely, direct lightning overvoltage; the other is the ground near the lightning line, which is caused by electromagnetic induction, namely, induced lightning overvoltage.

The lightning protection performance of transmission lines is mainly measured by lightning resistance level and lightning tripping rates. The maximum lightning current amplitude of insulation of transmission line that does not flashover when lightning strikes is called "the lightning resistance level" (the unit is kA). The lightning current which is lower than the lightning resistance level will not cause flashover when strike on the line. Otherwise, the flashover will inevitably occur. The number of trip on the transmission line per 100 km by lightning stroke is called "lightning tripping rate", which is a comprehensive index to implying the lightning protection performance of transmission line.

The lightning protection of transmission lines includes two periods: inductive lightning protection (the mid-1930s) and direct lightning protection. With the increasing transmission voltage, the direct lightning strike has become the main contradiction in lightning protection of transmission lines. At the beginning of the 1960s, the Monte Carlo method (a method of considering probability and statistic law) and computer were applied to the lightning protection design of the transmission line, which made the research results of the lightning protection be more in line with the engineering practice.

In determining the lightning protection mode of transmission lines, the importance of the transmission lines, the operation mode of the system, the intensity of the lightning activity on the transmission lines passing through the area, the characteristics of the terrain and foreign trade, the soil resistivity and so on should be taken into account. Combined with the operation experience of the original transmission lines, reasonable protection measures should be taken according to the results of technical and economic comparison.

(1) Erection of lightning lines

The most basic lightning protection measures for HV EHV and UHV transmission lines are lightning protection lines which are mainly to prevent direct lightning strikes. In addition, the lightning protection lines also have a shunt effect on lightning current, which can reduce the lightning current flowing into the tower and decrease the potential of the tower.

According to the regulations, 330 kV should be installed lightning protection lines on the whole line; and the 220 kV line shall be equipped with lightning protection lines generally; the 110 kV line should also be equipped with lightning protection line, on the whole line. However, it is not necessary to set up lightning protection lines along the whole line in areas with less slight lightning activity areas. The protection angle is usually 20°-30°. For the 220 kV and 330 kV double lightning protection line, the protection angle is about 20°.

In order to reduce the additional loss caused by the current in the lightning protection line under normal operation and use the lightning protection line as communication, the lightning protection line can be insulated through the small gap to the ground, the small gap is broken down and the lightning protection line is grounded during the lightning strike.

(2) Reducing the grounding resistance of tower

For the general height tower, reducing the grounding resistance of the tower is also an effective measure to improve the lightning resistance level of the line and prevent counterattacks. Each tower (power frequency grounding resistance without lightning wires) shall not exceed the value shown in table 6.3 when the lightning season.

In areas with low soil resistivity, the natural grounding resistance of the tower should be fully utilized. The method of extending the grounding line in parallel with the transmission line can reduce the voltage on the insulator string, so as to improve the lightning resistance level of the line.

Table 6.3   Power frequency grounding resistance of transmission line tower with lightning protection lines

Soil resistivity/Ω m	100 and below	100-500	500-1 000	1 000-2 000	2 000 and above
Grounding resistance/Ω	10	15	20	25	30

(3) Erection of coupled ground line

When it is difficult to reduce the grounding resistance of the tower, the method of laying ground lines under the transmission line can be adopted. The function of ground lines is to increase the coupling effect between the lightning arrester and to reduce the voltage on the insulator string. In addition, the coupled ground lines can increase the shunt effect of lightning current. The operation experience shows that the coupled ground wire can reduce the lightning tripping rate significantly.

(4) Use of unbalanced insulation

In modern HV EHV and UHV lines, the number of double circuit lines installed in the same rod is increasing gradually. When the usual lightning protection measures for this kind of line cannot meet the demand, the unbalanced insulation can be used to reduce the lightning tripping rates to keep the power supply going. The principle of unbalanced insulation is that the number of insulator strings in the two circuits is different. In this way, when lightning strikes, the circuit line with fewer insulator strings will have flashover, and the flashover conductor is equivalent to the ground wire. Thus, the lightning resistance level of the other circuit is improved. It is generally believed that the difference of insulation level between the two circuits should be $\sqrt{3}$ time of phase voltage (peak value), and if the difference is too large, the total failure rate will increase.

(5) Installation of automatic reclosing

Because most of the flashover caused by lightning can recover the insulation performance after tripping, the success rate of reclosing is high. According to statistics, the success rate of reclosing of 110 kV and above lines is about 75%-90%, and that for 35 kV and below lines is about 50%-80%. Therefore, all voltage lines should be installed automaticlly reclosing as far as possible.

(6) Adopt arc suppression coil grounding mode

For areas with strong lightning activity and the grounding resistance is difficult to reduce, the way of neutral ungrounded or grounded by the arc-suppression coil can be considered. Most of the one-way lightning flashover grounding faults can be eliminated by arc suppression coil. When two-phase or three-phase lightning strikes, the flashover of the first phase conductor caused by lightning does not cause tripping. The flashover line is equivalent to the ground line, which reduces the voltage on the insulator string of non flashover phase, thus improving the lightning resistance level.

(7) Installation of tubular arresters

Tubular arresters are usually installed at the crossing of the line and on the tower to limit overvoltage.

(8) Enhanced insulation

For the high tower, the lightning protection performance can be improved by increasing the number of insulator strings. The equivalent inductance of the high pole tower is large, the inductive overvoltage is large, and the winding failure rate increases with the height.

6.5.2 Overvoltage and protection in power system

In the power system, the overvoltage caused by its own internal reasons is called internal overvoltage. This is due to the presence of electric field inertial elements (capacitors) and magnetic field inertial elements (inductors) in power systems. When switching operations (including normal and accident operations), the power system will transition from one stable state to another state, and the electromagnetic field energy in the system components will be redistributed, which is an oscillating process. When the system parameters are mismatched, there will be a very high overvoltage.

Different from the single cause of lightning overvoltage (lightning discharge), the internal overvoltage has many kinds and different mechanisms because of its various causes, development process and influencing factors. Therefore, the internal overvoltage of power system includes two types, namely, steady-state over-voltage (namely temporary overvoltage) and operating over-voltage. As shown in Figure 6.13, it is a list of a number of internal overvoltages that occur frequently, have a great impact on the insulation, and have a typical development mechanism.

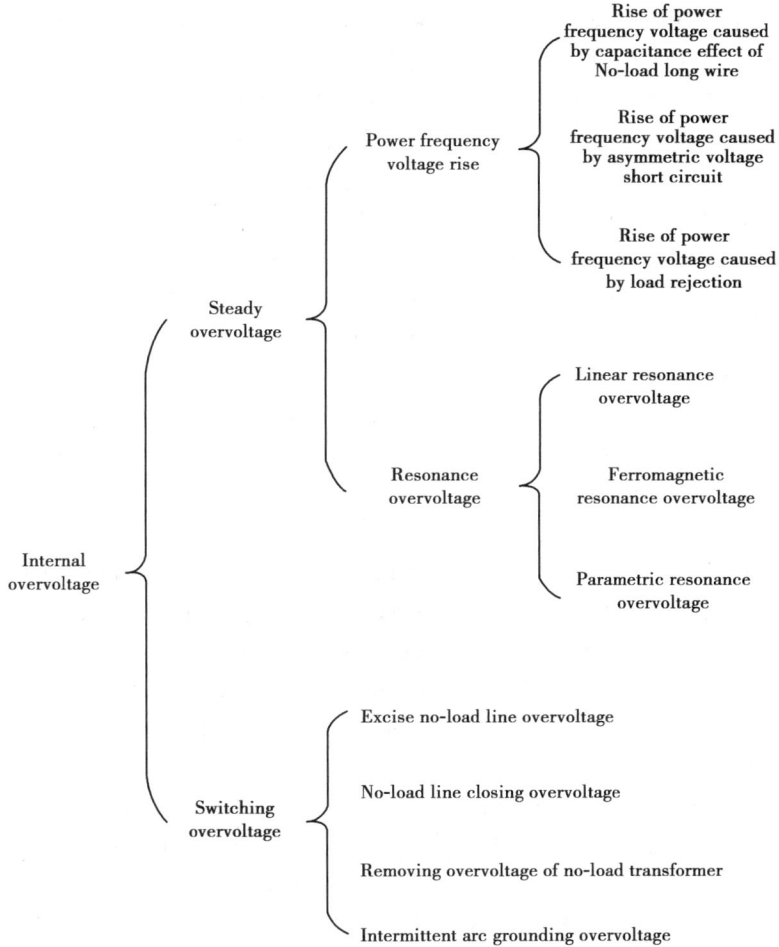

Figure 6.13　Types of internal overvoltage

Because the energy of the internal over-voltage comes from the power grid itself, its amplitude has a certain multiple relationship with the power frequency voltage of the power grid. In general, the amplitude um of internal overvoltage is expressed as a multiple of the maximum operating phase voltage amplitude (p. u.) of the system, namely, $U_m = K \cdot p. u.$.

Comparing with the absolute value of the overvoltage (kV) used in the case of lighting overvoltage, it is customary to use this multiple of overvoltage to indicate the magnitude of the internal overvoltage.

Although the duration of the internal overvoltage is much longer than that of the lighting overvoltage (the duration is only tens of μs), the duration of the general operating overvoltage is within 0. 1 s (5 power frequency cycles), and the duration of the temporary overvoltage is much longer.

6.5.3　Insulation coordination of power systems

The fundamental task of insulation coordination in power system is to correctly deal with the contradiction between overvoltage and insulation in order to achieve the high quality, safety and economical power supply. Insulation coordination is to determine the insulation level (insulation with-stand strength) of electrical equipment according to the limiting measures of various voltages and overvoltage that electrical equipment may bear in the system and the insulation performance of the equipment, so as to reduce the insulation damage caused by various voltages (normal working voltage and overvoltage).

Insulation coordination should not only deal with the coordination among various voltages, voltage limiting measures and insulation resistance of equipment, but also coordinate the relationship among investment cost, maintenance cost and accident loss cost economically. In addition, because of all kinds of overvoltage that may occur in the system, and the randomness of the insulation performance of electrical equipment and voltage limiting and protection equipment, insulation coordination is a complex problem and cannot be determined in isolation and simply according to a certain situation.

The core problem of insulation coordination is to determine the insulation level of electrical equipment. It is the priority of insulation design and often implied by the voltage test, since any kind of electrical equipment is not isolated in operation. Firstly, the electrical equipment must run with certain overvoltage protection devices; Secondly, the insulation of various electrical equipment and even the protection devices have mutual influence in operation, so when choosing the insulation level, we need to consider a variety of factors comprehensively. In fact, the insulation level is determined by the most stringent one of the three factors: the maximum working voltage for a long time, the atmospheric over-voltage and the internal overvoltage.

In 220 kV and below systems, it is uneconomical to limit the lighting overvoltage to lower than the internal overvoltage. Therefore, the insulation level of electrical equipment is mainly determined by the atmospheric overvoltage in these systems.

In UHV system, the amplitude of switching overvoltage increases with the voltage level and becomes the main contradiction gradually. However, the principle of internal overvoltage protection in UHV system is mainly to limit the operating overvoltage to a certain level by improving the performance of the circuit breaker (such as using the switch with parallel resistance); to limit the power frequency voltage to a certain level by the shunt reactor, and then to use the arrester as the backup protection of internal overvoltage. In this way, frequent actions will not occur under internal overvoltage. Because the internal overvoltage is limited to a certain level, the system insulation level is still determined by atmospheric overvoltage.

In the power grid of the polluted area, due to the influence of pollution, the insulation performance of the equipment will be greatly reduced. The pollution flashover accident often occurs in the normal working voltage under the bad weather conditions. The pollution flashover accident has a longer time, a wide spread and a great harm. In recent years, statistics show that the loss of pollution flashover accidents is more than the loss of lightning accidents, so the external insulation level of power grid in the seriously polluted area is mainly determined by the maximum operating voltage of the system.

The insulation level of an electrical equipment at a voltage level is the test voltage standard that the equipment can withstand (without flashover, breakdown or other damage). Considering that the electrical equipment must bear the action of operation voltage, power frequency voltage, atmospheric overvoltage and internal overvoltage during operation, the power frequency test voltage (1min) of various electrical equipment insulation (dry and wet lightning voltage are also specified for external insulation), lightning impulse test voltage and operation impulse test voltage are respectively specified in the test voltage standard. Considering the aging of insulation and the filthy performance of external insulation under the action of operating voltage and power frequency overvoltage, the long-time power frequency test voltage of some equipment is also specified.

For electrical equipment below voltage grade of 220 kV, 1min power frequency withstand voltage test is often used instead of lightning impulse test and operation impulse voltage test. The use of 1min power frequency test voltage to replace operating overvoltage and atmospheric overvoltage is due to that the operation time of the power frequency test voltage is longer, and the examination of the equipment is stricter. Moreover, the procedure of power frequency voltage withstand test is shown in the Figure 6.14.

Figure 6.14   the procedure of power frequency voltage withstand test

Where, β_1 is the lightning impulse coefficient (the ratio of lightning impulse withstand voltage to equivalent power frequency withstand voltage); β_2 is the operating impulse coefficient (operating impulse withstand voltage to equivalent power frequency withstand voltage).

The power frequency test voltage actually represents the total withstand level of insulation on internal and external overvoltage. Generally, in addition to the impulse withstand voltage test, as long as the equipment can pass the power frequency withstand voltage test, it is considered that the internal and external over-voltage can ensure the safety in operation.

It should be pointed out that, for EHV/UHV electrical equipment, it is generally considered that it is not appropriate to use power frequency withstand voltage test instead of operation impact withstand voltage test. Firstly, for the EHV/UHV, using 1 min power frequency test voltage instead of operating over-voltage may require insulation higher; secondly, the operation wave has a special effect on the insulation, and the voltage distribution inside the insulation is different from that under the power frequency voltage. Therefore, the operating wave test voltage is also specified for EHV electrical equipment.

The following methods of insulation coordination are used to determine the insulation level of electrical equipment:

(1) Conventional method

Up to now, the most widely used insulation coordination method is still the conventional method. Except for the statistical method used in the design of 330 kV and above (all are self-recovery insulation), the main method is still the conventional method in other cases.

The basic matching principle of conventional method is that all kinds of insulation are protected by the arrester, that is to say, the protection level of the arrester is the basis of determining the insulation level of the electrical equipment. According to international standards, the maximum voltage in residual voltage U_r, standard lightning impulse (1. 2/50 μs) discharge voltage $U_b(i)$ and the value of steep wave discharge voltage divided by 1. 15, is the protection level of the arrester.

(2) Statistical method

With the development of EHV transmission technology, the economic benefit by reducing the insulation level is more and more significant. In the conventional method, the lowest limit of insulation strength should be matched with the highest limit of overvoltage (maximum overvoltage value), and a sufficient safety margin should be also reserved. In fact, both of the overvoltage and the insulation strength are random variables, and their upper and lower limits cannot be calculated accurately. Moreover, the safety margin (matching coefficient or commonly used safety factor) selected by experience has certain randomness. These practices are unreasonable from an economic perspective, especially for EHV and UHV transmission systems. Therefore, it is necessary to consider the economic loss caused by insulation fault comprehensively, and then a reasonable conclusion can be drawn. In terms of comprehensive economic indicators, it is more reasonable to allow a certain insulation failure rate.

Due to the above reasons, since the 1960s, the international community began to explore the idea of insulation cooperation and formed the "Statistical method" gradually. IEC made a formal recommendation in the early 1970s, and now it has been used in the design of UHV external insulation in some countries. The premise of using a statistical method for insulation coordination is to fully grasp the statistical characteristics (probability density, distribution function, etc.) of various overvoltage and insulation strength as random variables.

Set the probability density function of overvoltage amplitude as $f(U)$, the probability function of insulation breakdown (or flashover) as $P(U)$. $f(U)$ is not related to $P(U)$, as shown in Figure 6. 15. $f(U_0)du$ is the probability of overvoltage occurring in the range of du around U_0, while $P(U_0)$ is the probability of insulation breakdown under the action of voltage U_0. Because they are independent of each other, the probability of insulation breakdown is

$$P(U_0)f(U_0)\,\mathrm{d}U = \mathrm{d}R \tag{6.3}$$

where $\mathrm{d}R$ is called differential failure rate, that is, the small area in the figure with oblique shadow.

When calculating the overvoltage in the power system, it usually only uses the absolute value, regardless of the polarity (we can think that the positive and the negative polarity are half of each). According to the definition, the distribution range of overvoltage amplitude should be U_φ-∞ (maximum operating phase voltage amplitude), so the insulation failure rate is

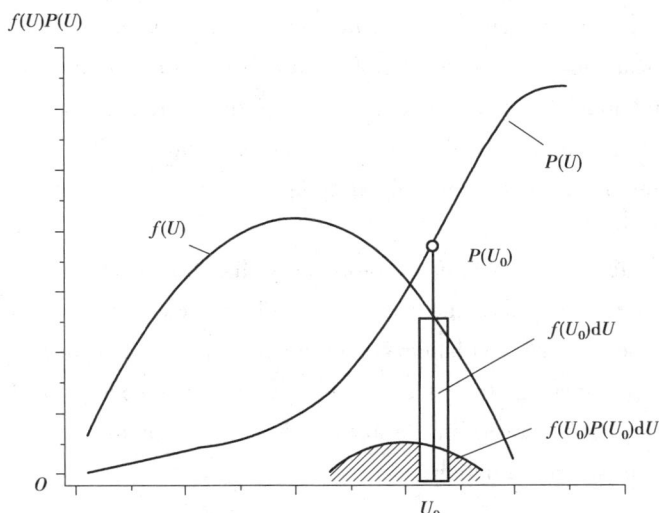

Figure. 6.15　Insulation failure rate

$$R = \int_{U_\varphi}^{\infty} P(U)f(U)\,\mathrm{d}U \tag{6.4}$$

That is the total area of shadow in Figure 6.7. It is the probability of breakdown (or flashover) of the insulation under overvoltage.

If the electrical strength of the insulation is increased, the $P(U)$ curve in Figure 6.7 moves to the right. The area of the shadow part is reduced. The failure rate of the insulation is reduced. However, the equipment investment will increase. Thus, we can adjust certain factors according to the need by using statistical methods. For example, according to the requirements of optimizing the overall economic index, the coordination between the insulation cost and the accident loss is carried out, and the reasonable insulation level is selected on the premise of meeting the predetermined insulation failure rate. When using a statistical method for insulation coordination, safety margin is no longer a random value, but a variable related to the insulation failure rate.

It is not difficult to see that it is rather complicated and difficult to use the above statistical method to carry out insulation coordination in practical engineering. For this reason, IEC also recommends a "simplified statistical method" for practical application.

In the simplified statistical method, some assumptions are made for the statistical concerns of overvoltage and insulation strength, such as assuming that they all follow normal distributions and their standard deviations are known. In this way, their probability distribution curves can be represented by points corresponding to a certain reference probability, which are respectively called "statistical overvoltage U_s" (reference cumulative probability is 20%) and "statistical insulation withstand voltage U_w" (reference withstand probability is 90%, namely, breakdown probability is 10%). They are also linked by a coefficient called "statistical safety factor".

$$K_S = \frac{U_w}{U_s} \tag{6.5}$$

When the overvoltage remains constant, the statistical insulation voltage and the statistical safety factor will increase and the insulation failure rate will decrease with the improvement of insulation level. The simplified statistical method is essentially a hybrid insulation coordination method, which utilizes the probability and statistical characteristics of the concerned parameters. Applying this method to the self-healing insulation with known characteristic probability, the insulation failure rate R can be calculated under different statistical safety factors (K_S), which is important for evaluating the operating reliability of the system.

In fact, it is very difficult to obtain the probability distribution of non self-restoring insulation breakdown voltage. The reason is that one sample can only provide one data. Therefore, the conventional method is still used in the insulation coordination of non self-recovery insulation of various voltage levels; the conventional method is also used for the self-recovery insulation of 220 kV and below. Only for the UHV self-recovery insulation (such as line insulation), the simplified statistical method is used for insulation coordination.

Chapter 7

Smart Grid

A smart grid is an electrical grid that includes a variety of operational and energy measures including smart meters, smart appliances, renewable energy resources, and energy efficient resources. Electronic power conditioning and control of the production and distribution of electricity are important aspects of the smart grid.

Roll-out of smart grid technology also implies a fundamental re-engineering of the electricity services industry, although typical usage of the term is focused on the technical infrastructure.

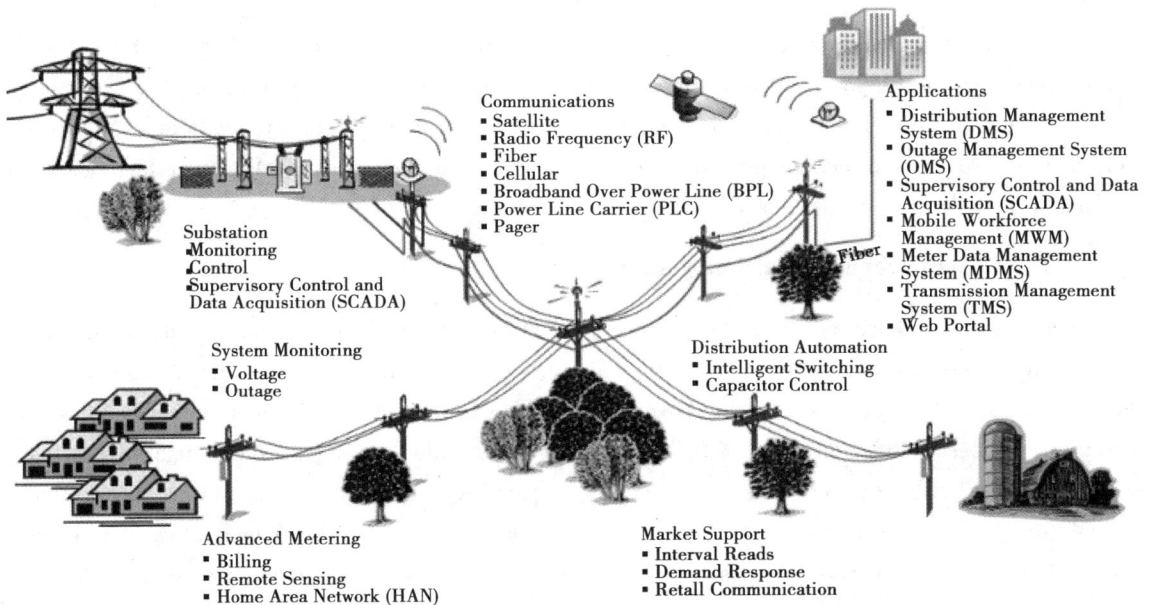

Figure 7.1   Smart grid technology

7.1　Background

Since the early 21st century, opportunities to take advantage of improvements in electronic communication technology to resolve the limitations and costs of the electrical grid have become apparent. Technological limitations on metering no longer force peak power prices to be averaged out and passed on to all consumers equally. At the same time, growing concerns over environmental damage from fossil-fired power stations have led to a desire to use large amounts of renewable energy. Dominant forms such as wind power and solar power, are highly variable, and so the need for more sophisticated control systems became apparent to facilitate the connection of sources to the other highly controllable grid. Power from photovoltaic cells (and to a lesser extent, wind turbines) has also significantly called into question the imperative for large, centralized power stations. The rapidly falling costs point to a major change from the centralized grid topology to one that is highly distributed, with power being both generated and consumed right at the limits of the grid. Finally, growing concern over the terrorist attack in some countries has led to calls for a more robust energy grid that is less dependent on centralized power stations that could be perceived to be potential attack targets.

7.1.1　Definition of "smart grid"

The smart grid is the intelligent power grid (smart power), also known as "grid 2.0". Based on the integrated and high-speed two-way communication network, the reliability, safety, economy, efficiency and environmental friendliness of power grid is realized via the application of advanced sensor and measurement technology, advanced equipment technology, advanced control methods and advanced decision support system technology.

7.1.2　Early technological innovations

Smart grid technologies emerged from earlier attempts at using electronic control, metering, and monitoring. In the 1980s, automatic meter reading was used for monitoring loads from large customers, and in the 1990s evolved into the Advanced Metering Infrastructure with meters that could store electricity useage at different times of the day. Smart meters add continuous communications so that monitoring can be done in real time and be used as a gateway to demand response-aware devices and "smart sockets" in the home. Early forms of such demand-side management technologies were dynamic demand aware devices that passively sensed the load on the grid by monitoring changes in the power supply frequency. Devices such as industrial and domestic air conditioners, refrigerators and heaters adjusted their duty cycle to avoid activation during times when the grid was suffering a peak condition. Beginning in 2000, Italy's Telegestore Project was the first network large numbers (27 million) of homes using smart meters connected via low bandwidth power line communication. Some experiments used the term broadband over power lines (BPL), while others used wireless technologies such as mesh networking promoted for more reliable connections to disparate devices at home as well as supporting metering of other utilities such as gas and water.

Monitoring and synchronization of wide-area networks were revolutionized in the early 1990s when the Bonneville Power Administration expanded its smart grid research with prototype sensors that were capable of very rapid analysis of anomalies in electricity quality over very large geographic areas. The culmination of this work was the first operational Wide Area Measurement System (WAMS) in 2000. Other countries are rapidly integrating this technology — China started to have a comprehensive national WAMS when the past 5-year economic plan was completed in 2012.

The earliest deployments of smart grids include the Italian system Telegestore (2005), the mesh network of Austin, Texas (since 2003), and the smart grid in Boulder, Colorado (2008).

7.2　Features of the smart grid

The smart grid represents the full suite of current and proposed responses to the challenges of electricity supply. Because of the diverse range of factors, there are numerous competing taxonomies and no agreement on a universal definition. Nevertheless, one possible categorization is presented here.

7.2.1　Reliability

The smart grid makes use of technologies such as state estimation to improve fault detection and allow self-healing of the network without the intervention of technicians. This will ensure a more reliable supply of electricity and reduce vulnerability to natural disasters or attacks.

Although multiple routes are touted as a feature of the smart grid, the old grid also featured multiple routes. Initial power lines in the grid were built using a radial model, later connectivity was guaranteed via multiple routes, referred to as a network structure. However, this created a new problem: if the current flow or related effects across the network exceed the limits of any particular network element, it could fail, and the current would be shunted to other network elements, which eventually may fail also, causing a domino effect. A technique to prevent this is load shedding by rolling blackout or voltage reduction (brownout).

The economic impact of improved grid reliability and resilience is the subject of a number of studies and can be calculated using a US-DOE funded methodology for US locations using at least one calculation tool.

7.2.2　Flexibility in network topology

Next-generation transmission and distribution infrastructure will be better able to handle possible bidirectional energy flows, allowing for distributed generation such as from photovoltaic panels on building roofs, but also the use of fuel cells, charging to/from the batteries of electric cars, wind turbines, pumped hydroelectric power and other sources.

Classic grids were designed for a one-way flow of electricity, but if a local sub-network gener-

117

ates more power than it is consuming, the reverse flow can raise safety and reliability. A smart grid aims to manage these situations.

7.2.3 Efficiency

Numerous contributions to an overall improvement of the efficiency of energy infrastructure are anticipated from the deployment of smart grid technology, in particular, including demand-side management, for example, turning off air conditioners during short-term spikes in electricity price, reducing the voltage when possible on distribution lines through Voltage/VAR Optimization (VVO), eliminating truck-rolls for meter reading, and reducing truck-rolls by improved outage management using data from Advanced Metering Infrastructure systems. The overall effect of less redundancy in transmission and distribution lines, and greater utilization of generators leads to lower power prices.

7.2.4 Load adjustment/Load balancing

The total load connected to the power grid can vary significantly over time. Although the total load is the sum of many individual choices of the clients, the overall load is not necessarily stable or slow varying. For example, if a popular television program starts, millions of televisions will start to draw current instantly. Traditionally, to respond to a rapid increase in power consumption, faster than the start-up time of a large generator, some spare generators are put on a dissipative standby mode(citation needed). A smart grid may warn all individual television sets or another larger customer, to reduce the load temporarily (to allow time to start up a larger generator or continuously in the case of limited resources). Using mathematical prediction algorithms it is possible to predict how many standby generators need to be used to reach a certain failure rate. In the traditional grid, the failure rate can only be reduced at the cost of more standby generators. In a smart grid, the load reduction by even a small portion of the clients may eliminate the problem.

While traditionally load balancing strategies aim to change consumers' consumption patterns to make demand more uniform, developments of energy storage and individual renewable energy generation have provided opportunities to devise balanced power grids without affecting consumers' behavior. Typically, storing energy during off-peak times eases the high demand of supply during peak hours. Dynamic game-theoretic frameworks have proved particularly efficient at storage scheduling by optimizing energy cost using their Nash equilibrium.

7.2.5 Sustainability

The improved flexibility of the smart grid permits greater penetration of highly variable renewable energy sources such as solar power and wind power, even without the addition of energy storage. Current network infrastructure can not accomodate for many distributed feed-in points, and even if some feed-in is allowed at the local (distribution) level, the transmission-level infrastructure cannot accommodate it. Rapid fluctuations in distributed generation, such as due to cloudy or gusty weather, present significant challenges to power engineers to ensure stable power levels through varying the output of the more controllable generators such as gas turbines and hydroelectric generators.

Smart grid technology is a necessary condition for very large amounts of renewable electricity on the grid for this reason.

7.2.6 Demand response support

Demand response support allows generators and loads to interact in an automated fashion in real time, coordinating demand to flatten spikes. Eliminating the fraction of demand that occurs in these spikes eliminates the cost of adding reserve generators, cuts wear and tear and extends the life of the equipment, and allows users to cut their energy bills by telling low priority devices to use energy only when it is cheapest.

Currently, power grid systems have varying degrees of communication within control systems for their high-value assets, such as in generating plants, transmission lines, substations and major energy users. In general, information flows one way, from the users and the loads they control back to the utilities. The utilities attempt to meet the demand and succeed or fail to varying degrees (brownouts, rolling blackout, uncontrolled blackout). The total amount of power demand by the users can have a very wide probability distribution which requires spare generating plants in standby mode to respond to the rapidly changing power usage. This one-way flow of information is expensive; the last 10% of generating capacity may be required as little as 1% of the time, and brownouts and outages can be costly to consumers.

Demand response can be provided by commercial, residential loads, and industrial loads. For example, Alcoa's Warrick Operation is participating in MISO as a qualified Demand Response Resource, and the Trimet Aluminum uses its smelter as a short-term mega-battery.

Latency of the data flow is a major concern, with some early smart meter architectures allowing actually as long as 24 hours delay in receiving the data, preventing any possible reaction by either supplying or demanding devices.

7.2.7 Platform for advanced services

The use of robust two-way communications, advanced sensors, and distributed computing technology will improve the efficiency, reliability and safety of power delivery and use. It also opens up the potential for entirely new services or improvements on existing ones, such as fire monitoring and alarms that can shut off power, make phone calls to emergency services, etc.

7.3 Smart grid technology

The bulk of smart grid technologies are already used in other applications such as manufacturing and telecommunications and are being adapted for use in grid operations.

Improving areas for integrated communications include: substation automation, demand response, distribution automation, supervisory control and data acquisition (SCADA), energy management systems, wireless mesh networks and other technologies, power-line carrier communications, and fiber-optics. Integrated communications will allow for real-time control, information and data exchange to optimize system reliability, asset utilization, and security.

Sensing and measurement: core duties are evaluating congestion and grid stability, monitoring equipment health, energy theft prevention, and control strategies support. Technologies include advanced microprocessor meters (smart meter) and meter reading equipment, wide-area monitoring systems, dynamic line rating (typically based on online readings by Distributed temperature sensing combined with Real time thermal rating (RTTR) systems), electromagnetic signature measurement/ analysis, time-of-use and real-time pricing tools, advanced switches and cables, backscatter radio technology, and Digital protective relays.

Distributed power flow control: power flow control devices clamp onto existing transmission lines to control the flow of power within. Transmission lines enabled with such devices support greater use of renewable energy by providing more consistent, real-time control over how that energy is routed within the grid. This technology enables the grid to effectively store more intermittent energy from renewables for later use.

Smart power generation is a concept of matching electricity generation with demand using multiple identical generators which can start, stop and operate efficiently at chosen load, independently of the others, making them suitable for base load and peaking power generation. Matching supply and demand, called load balancing, is essential for a stable and reliable supply of electricity. Short-term deviations in the balance lead to frequency variations and prolonged mismatch results in blackouts. Operators of power transmission systems are charged with the balancing task, matching the power output of all the generators to the load of their electrical grid. The load balancing task has become much more challenging when increasingly intermittent and variable generators such as wind turbines and solar cells are added to the grid, forcing other producers to adapt their output much more frequently than that in the past.

Power system automation enables rapid diagnosis of and precise solutions to specific grid disruptions or outages. These technologies rely on and contribute to each of the other four key areas. Three technology categories for advanced control methods are distributed intelligent agents (control systems), analytical tools (software algorithms and high-speed computers), and operational applications (SCADA, substation automation, demand response, etc.). Using artificial intelligence programming techniques, the Fujian power grid in China created a wide area protection system is able to accurately calculate a control strategy and execute it rapidly. The Voltage Stability Monitoring & Control (VSMC) software uses a sensitivity-based successive linear programming method to reliably determine the optimal control solution.

7.4　Smart grid modelling

Many different concepts have been used to model intelligent power grids. They are generally studied within the framework of complex systems. In a recent brainstorming session, the power grid was considered within the context of optimal control, ecology, human cognition, glassy dynamics, information theory, microphysics of clouds, and many others. Here is a selection of the types of analyses that have appeared in recent years.

The self-validated and superised protection system

Pelqim Spahiu and Ian R. Evans introduced the concept of a substation-based smart protection and hybrid Inspection Unit in their study.

Kuramoto oscillators

The Kuramoto model is a well-studied system. The power grid has been described in this context as well. The goal is to keep the system in balance or to maintain phase synchronization (also known as phase locking). Non-uniform oscillators also help to model different technologies, different types of power generators, patterns of consumption, and so on. The model has also been used to describe the synchronization patterns in the blinking of fireflies.

Bio-systems

Power grids have been related to complex biological systems in many other contexts. In one study, power grids were compared to the dolphin social network. These creatures streamline or intensify communication in case of special situations. The intercommunications that enable them to survive are highly complex.

Random fuse networks

In the percolation theory, random fuse networks have been studied. The current density might be too low in some areas, and too strong in others. Therefore, the analysis can be used to smooth out potential problems in the network. For instance, high-speed computer analysis can predict blown fuses and correct for them, or analyze patterns that might lead to a power outage. It is difficult for humans to predict the long term patterns in complex networks, so fuse or diode networks are used instead.

Smart Grid Communication Network

Network Simulators are used to simulate/emulate network communication effects, typically involving setting up a lab with the smart grid devices, applications etc. with the virtual network being provided by the network simulator.

Neural networks

Neural networks have been considered for power grid management as well. Electric power systems can be classified in multiple different ways: non-linear, dynamic, discrete, or random. Artificial Neural Networks (ANNs) attempt to solve the most difficult of these problems, the non-linear problems.

Demand Forecasting

One application of ANNs is in demand forecasting. For grids to operate economically and reliably, demand forecasting is essential, because it is used to predict the amount of power that will be consumed by the load. This is dependent on weather conditions, type of day, random events, incidents, etc. For non-linear loads though, the load profile isn't smooth and as predictable, resulting in higher uncertainty and less accuracy using the traditional Artificial Intelligence models. Some factors that ANNs consider when developing these sort of models: classification of load profiles of different customer classes based on the consumption of electricity, increased responsiveness of demand to predict real time electricity prices as compared to conventional grids, the need to input past demand as different components, such as peak load, base load, valley load, average load, etc. instead of joining them into a single input, and lastly, the dependence of the type on specific input variables. An example of the last case would be given due to the type of day, whether on weekday or weekend, that has little impact on Hospital grids, but has a significant impact on the resident housing grids' load profile.

Markov processes

As wind power continues to gain popularity, it becomes a necessary ingredient in realistic power grid studies. Off-line storage, wind variability, supply, demand, pricing, and other factors can be modeled as a mathematical game. Here the goal is to develop a winning strategy. Markov processes have been used to model and study this type of system.

Maximum entropy

All of these methods are, in one way or another, maximum entropy methods. They have is an active area of research. This goes back to the ideas of Shannon, and many other researchers who studied communication networks. Continuing along similar lines today, modern wireless network research often considers the problem of network congestion, and many algorithms are being proposed to minimize it, including game theory, innovative combinations of FDMA, TDMA, and others.

7.5　Deployments and potential deployments

Certain deployments utilize the OpenADR standard for load shedding and demand reduction during higher demand periods.

7.5.1　China

The smart grid market in China is estimated to be $22.3 billion with a projected growth to $61.4 billion by 2015. Honeywell is developing a demand response pilot and feasibility study for China with the State Grid Corp. of China using the OpenADR demand response standard. The State Grid Corp., the Chinese Academy of Science, and General Electric intend to work together to develop standards for China's smart grid rollout.

7.5.2　United Kingdom

The OpenADR standard was demonstrated in Bracknell, England, where peak use in commercial buildings was reduced by 45 percent. As a result of the pilot, the Scottish and Southern Energy (SSE) said it would connect up to 30 commercial and industrial buildings in Thames Valley, west of London, to a demand response program.

7.5.3　**United States**

In 2009, the US Department of Energy awarded an $11 million grant to Southern California Edison and Honeywell for a demand response program that automatically turned down energy use during peak hours for participating industrial customers. The Department of Energy awarded an $11.4million grant to Honeywell to implement the program using the OpenADR standard.

Hawaiian Electric Co. (HECO) is implementing a two-year pilot project to test the ability of an ADR program to respond to the intermittence of wind power. Hawaii has a goal to obtain 70 percent of its power from renewable sources by 2030. HECO will give customers incentives for reducing power consumption within 10 minutes of notice.

Appendix *1*

Major Journal in the Electrical Engineering

[1] Proceedings of the IEEE

[2] IEEE Signal Processing Magazing

[3] IEEE Transactions on Energy Conversion

[4] IEEE Transactions on Power Systems

[5] IEEE Transactions on Power Electonics

[6] IEEE Transactions on Power Delivery

[7] IEEE Transactions on Dielectrics and Electrical Insulationg

[8] IEEE Transactions on Industry Applications

[9] IEEE Transactions on Industrial Electronics

[10] IEEE Transactions on Smart Grid

[11] IEEE Transactions on Sustainable Energy

[12] IEEE Transactions on Magnetics

[13] IEEE Transactions on ELectromagnetic Compatibility

[14] IEEE Transactions on Applied Superconductivity

[15] Proceedings of the CSEE

[16] Transactions of China Electrotechnical Society

[17] IET Power Electronics

[18] IET Signal Processing

[19] IET Generation, Transmission & Distribution

[20] IET Electric Power Applications

[21] IET Circuits, Devices & Systems

[22] IET Renewable Power Generation

[23] IET Electric Power Applications

[24] International Transactions on Electrical Energy Systems

Appendix **2**

Basic Definition Terminology

Current: The flow of electricity or the movement of electrons through a conductor typically measured in ampere.

Voltage: Electric "pressure" measured in volts. Power systems are typically measured in 1 000 s volts or kV.

Watt: Unit of electrical power. 1 MW is one million watts.

Power: Rate at which electricity does work. Measured in watts or kilowatts (kW) or megawatts (MW).

Energy: The amount of work that can be done by electricity, typically measured in kilowatt-hours (kWh) or megawatthours (MWh).

Alternating Current (AC): Electric current in which the direction of the current's flow is reversed or alternated at 50 Hz in the China.

Direct Current (DC): Electric current flows continuously in the same direction as contrasted with alternating current.

Generation: The production of electric energy. Fossil fuels, wind turbines, solar panels, and other technologies are used to generate electricity.

Electric Load: Electricity consumers, such as residences, businesses, and government centers that use electricity.

Load Center: A particular geographical area where energy is used. Most commonly refers to an area within a utility's service territory where energy demand is highest (i. e. , cities, major industrial areas, etc.).

Electric Power Transmission: The process by which large amounts of electricity produced are transported over long distances for eventual use by consumers.

Substation: A part of an electrical transmission system that transforms voltage from high to low, or the reverse.

Interconnection: Points on a grid or network where two or more transmission lines are connected at a substation or switching station, or where one stage of the energy supply chain meets the next.

Transmission Line: A line that carries electricity at voltages of 69 kV or greater and is used to transmit electric power over relatively long distances, usually from a central generating station to main substations.

Distribution Line: A line that carries electricity at lower voltages of 10 kV to 35 kV and is used to distribute power drawn from high-voltage transmission systems to end-use customers.

Conductors (Power Lines): Metal cables used for carrying electric current.

Foundation: System that transfers to the ground the various dead and live loads of the transmission structure and conductors.

Insulators: Used to contain, separate, or support electrical conductors.

Shield and Ground Wire: Wires used primarily for protection from lightning strikes and corresponding surges.

Transmission Structures: Used to keep high-voltage conductors (power lines) separated from their surroundings and from each other.

Corona: Electrical breakdown of the air near high voltage conductors into charged particles.

References

[1] ALEXANDER C K, SADIKU M N O. Fundamentals of Electric Circuits[M]. New York: McGraw-Hill, 2013.

[2] RASHID M H. Power Electronics: Circuits, Devices and Applications[M]. Englewood: Prentice Hall, 2013.

[3] HAMBLEY A R. Electrical Engineering Principles and Applocations[M]. New York: Pearson Education, 2014.

[4] LAI C M, CHENG Y H, Ming-Hua Hsieh, et al. Development of a Bidirectional DC/DC Converter With Dual-Battery Energy Storage for Hybrid Electric Vehicle System[J]. IEEE Transactions on Vehicular Technology, 2018, 67(2):1036-1052.

[5] CHAUDHURI N R. Integrating Wind Energy to Weak Power Grids using High Voltage Direct Current Technology[M]. Berlin: Springer Press, 2018.

[6] QIU P, HUANG X M, Wang Yi, et al. Application of High Voltage DC Circuit Breaker in Zhoushan VSC-HVDC Transmission Project[J]. High Voltage Engineering, 2019, 44(2): 403-408.

[7] DENNO K. High Voltage Engineering in Power Systems[M]. Baca Raton: CRC Press, 2018.

[8] QU B J, YANG Q X, LI Y J, et al. A New Concentration Detection System for SF_6/N_2 Mixture Gas in Extra/Ultra High Voltage Power Transmission Systems[J]. IEEE Sensors Journal, 2018, 18(9):3806-3812.

[9] HAUSCHILD W, LEMKE E. High-Voltage Test and Measuring Techniques[M]. Berlin: Springer-Verlag, 2014.

[10] KUFFEL E, ZAENGL W S, KUFFEL J. High Voltage Engineering Fundamentals[M]. Amsterdam: Elsevier, 2000.

[11] CHEN S Y, SONG S F, LI L X, et al. Survey on Smart Grid Technology[J]. Power System Technology, 2009, 33(8):1-7.

［12］ BUSH S F. Smart Grid: Communication-Enabled Intelligence for the Electric Power Grid
［M］. Hoboken: Wiley-IEEE Press, 2015.

［13］ BORLASE S. Smart Grid: Infrastructure, Technology, and Solutions［M］. Boca Raton: CRC
Press, 2015.

［14］ BABONNEAU F L F, HAURIE A. Energy Technology Environment Model with Smart Grid
and Robust Nodal Electricity Prices［J］. Annals of Operations Research, 2019, 274(2):
1-17.